中国地质调查成果 CGS 2019-022
中国地质调查局 DD20190603，DD20190373，
DD20190361，DD20160038 项目科普成果

STROMATOLITES IN SHENNONGJIA
THE MICROBES BUILT GREAT BARRIER REEFS MORE THAN 1000 MILLION YEARS AGO

QIAN MAIPING XING GUANGFU MA XUE LI XIAOCHI SUO YINGPING XIANG HONGLI PAN JIPEI

神农架叠层石
10多亿年前远古海洋微生物建造的大堡礁

钱迈平　邢光福　马　雪　李晓池　所颖萍　项红莉　潘继培　著

中国科学技术大学出版社
UNIVERSITY OF SCIENCE AND TECHNOLOGY OF CHINA PRESS

图书在版编目(CIP)数据

神农架叠层石:10多亿年前远古海洋微生物建造的大堡礁:汉、英/钱迈平等著.—合肥:中国科学技术大学出版社,2019.6
ISBN 978-7-312-04644-5

Ⅰ.神…　Ⅱ.钱…　Ⅲ.神农架—叠层石—汉、英　Ⅳ.Q914.82

中国版本图书馆CIP数据核字(2019)第040924号

出版	中国科学技术大学出版社 安徽省合肥市金寨路96号,230026 http://press.ustc.edu.cn https://zgkxjsdxcbs.tmall.com
印刷	合肥市宏基印刷有限公司
发行	中国科学技术大学出版社
经销	全国新华书店
开本	787 mm×1092 mm　1/16
印张	20.75
字数	412千
版次	2019年6月第1版
印次	2019年6月第1次印刷
定价	120.00元

内容简介

10多亿年前,神农架是一片汪洋大海,其中一些面积广大的浅滩潮坪,发育了许多由单细胞的蓝细菌等微生物建造的叠层石生物礁,甚至在一些地方构成了绵延可达数十至上百千米的大堡礁,规模宏大,形态多样,是当时地球生态系统重要的组成部分。它们通过光合作用制造自己生命活动所需的营养,消耗了大气中的二氧化碳,吸收了大气中的氮,释放出氧气,逐渐改造了地球的环境,为包括人类在内的多细胞生命的兴起,创造了必不可少且至关重要的条件。保存下来的化石遗迹也成为今天神农架众多壮丽景观的组成部分,其中许多体态奇异、花纹精美的叠层石化石也成为今天收藏观赏的奇石珍品。本书将以图文并茂的形式,展示这些远古海洋叠层石大堡礁的遗迹,讲述微生物建造叠层石在早期地球演化历史中所起的重大作用,以及生物与环境的紧密联系、相互影响、共同演化的过程,从而使我们更加深刻地理解珍爱生命、保护我们唯一家园——地球的重要性。

SUMMARY

More than 1,000 million years ago, Shennongjia was a vast expanse of sea distributed carbonate platforms developed huge and various stromatolite bioherms and biostromas built by microbes such as unicellular cyanobacteria etc. and some of bioherms and biostromas stretched several dozens or hundreds kilometers were almost comparable to modern Great Barrier Reef. They played an important role in palaeoecosystem through consuming carbon dioxide and nitrogen, meanwhile producing oxygen as a byproduct of their photosynthesis in the atmosphere of early Earth, and transformed gradually the global environments and created important indispensable conditions for the rise of complex multicellular organisms including ourselves. Now, these fossilized remains of the giant stromatolite bioherms and biostromas are constituent parts of many magnificent landscapes, while those exquisite and extraordinary specimens of fossilized stromatolites became currently precious collectibles. Here we want to intersperse this book with pictures to show the stromatolites in the Shennongjia, and tell about the microbes built stromatolites played a significant role in evolution of early Earth, while organisms and enviroments go hand in hand interactionally. Thus, we can profoundly understand the importance of loving life and protecting our only home — the Earth.

作者简介　　AUTHORS

钱迈平

1954年9月生，博士，南京大学地球科学系毕业。现为中国地质调查局南京地质调查中心研究员，江苏省古生物学会常务理事。长期从事华东、华北、西南和新疆地区地层学、古生物学研究和矿产资源调查。

Qian Maiping

Born in September 1954, PhD, Professor of stratigraphy and paleontology at Nanjing Center of China Geological Survey.

邢光福

1965年2月生，博士，南京大学地球科学系毕业。现任中国地质调查局南京地质调查中心总工程师、研究员，中国矿物岩石地球化学学会、中国灾害防御协会火山专业委员会、中国旅游地学研究会理事。长期从事华东地区、华南地区、西太平洋沿岸、南美及南极大地构造学研究和矿产资源调查。

Xing Guangfu

Born in February 1965, PhD, the Chief Engineer and Professor of mineralogy, petrology and geochemistry at Nanjing Center of China Geological Survey.

马雪

1984年3月生，硕士，成都理工大学沉积地质研究院毕业。现任中国地质调查局南京地质调查中心助理研究员，长期从事华东地区及青藏高原沉积学研究和矿产资源调查。

Ma Xue

Born in March 1984, MS, Research Assistant of sedimentology at Nanjing Center of China Geological Survey.

作者简介　　AUTHORS

李晓池

1945年6月生,博士,北京地质学院及新西兰奥克兰大学地质系毕业。现任中国神农架世界地质公园地质研究所所长,长期从事华中地区、华南地区及南非地质研究工作。

Li Xiaochi

Born in June 1945, PhD, the Director of The Institute of Geology at Shennongjia UNESCO Global Geopark of China.

所颖萍

1980年1月生,硕士,南京大学地球科学与工程学院毕业。现任中国地质调查局南京地质调查中心工程师,科技处副处长,江苏省地质学会副秘书长,长期从事地质科研管理工作。

Suo Yingping

Born in January 1980, MS, the Deputy Director for Science and Technology at Nanjing Center of China Geological Survey.

项红莉

1987年5月生,硕士,南京大学外国语学院毕业。现任中国地质调查局南京地质调查中心科技处外事管理、翻译,从事地质科研项目管理工作。

Xiang Hongli

Female, born in May 1987, MA, Manager of Foreign Affairs at Nanjing Center of China Geological Survey, Translator.

作者简介　　AUTHORS

潘继培

1976年10月生,大学本科,现任中国地质调查局南京地质调查中心江苏省黄金珠宝检测中心检测员,长期从事贵金属珠宝矿物检测工作。

Pan Jipei

Born in October 1976, Undergraduate, the noble metals and gems tester of Jiangsu Gold and Jewelry Testing Center at Nanjing Center of China Geological Survey.

目　录

0　引言 ··· 1
1　叠层石概述 ·· 7
　1.1　叠层石是什么？ ··· 8
　1.2　叠层石化石是我们直接用肉眼就可看到的地球最古老的生命记录 ·· 12
　1.3　叠层石与有氧大气圈及多细胞生物的兴起 ············ 31
　1.4　现代叠层石 ·· 52
　1.5　叠层石成矿作用 ··· 58
　1.6　叠层石研究历史和现状 ······································ 66
2　神农架地质公园主要景区的中元古代叠层石大堡礁群遗迹 ··· 73
　　2.1　乱石沟组 ··· 78

CONTENTS

0　INTRODUCTION ··· 1
1　BRIEF INTRODUCTION OF STROMATOLITES ········ 7
　1.1　WHAT ARE STROMATOLITES? ····················· 8
　1.2　STROMATOLITE FOSSILS ARE THE EARTH'S OLDEST LIFE RECORD VISIBLE TO OUR NAKED EYES ··· 12
　1.3　STROMATOLITES, OXYGENATED ATMOSPHERE AND THE RISE OF MULTICELLULAR ORGANISMS ··· 31
　1.4　MODERN STROMATOLITES ························· 52
　1.5　STROMATOLITE-HOSTED MINERALIZATIONS ··· 58
　1.6　HISTORY AND CURRENT STATUS OF STUDY ON STROMATOLITES ··· 66
2　REMAINS OF THE MESOPROTEROZOIC STROMATOLITE GREAT BARRIER REEFS IN MAIN TOURIST SPOTS OF SHENNONGJIA UNESCO GLOBAL GEOPARK ··· 73
　　2.1　LUANSHIGOU FORMATION ························ 78

2.2 大窝坑组 ⋯⋯⋯⋯⋯⋯⋯⋯⋯⋯⋯⋯⋯⋯⋯⋯⋯⋯⋯⋯⋯⋯ 124
2.3 矿石山组 ⋯⋯⋯⋯⋯⋯⋯⋯⋯⋯⋯⋯⋯⋯⋯⋯⋯⋯⋯⋯⋯⋯ 144
2.4 台子组 ⋯⋯⋯⋯⋯⋯⋯⋯⋯⋯⋯⋯⋯⋯⋯⋯⋯⋯⋯⋯⋯⋯⋯ 184
2.5 野马河组 ⋯⋯⋯⋯⋯⋯⋯⋯⋯⋯⋯⋯⋯⋯⋯⋯⋯⋯⋯⋯⋯⋯ 226
2.6 温水河组 ⋯⋯⋯⋯⋯⋯⋯⋯⋯⋯⋯⋯⋯⋯⋯⋯⋯⋯⋯⋯⋯⋯ 233
2.7 石槽河组 ⋯⋯⋯⋯⋯⋯⋯⋯⋯⋯⋯⋯⋯⋯⋯⋯⋯⋯⋯⋯⋯⋯ 240
2.8 神农架叠层石大堡礁的终结 ⋯⋯⋯⋯⋯⋯⋯⋯⋯⋯⋯⋯⋯⋯ 253

3 神农架中元古代叠层石 ⋯⋯⋯⋯⋯⋯⋯⋯⋯⋯⋯⋯⋯⋯⋯⋯⋯ **263**
3.1 神农架大圆顶叠层石（*Megadomia shennongjiaensis* Qian et al., 2017）⋯⋯⋯⋯⋯⋯⋯⋯⋯⋯⋯⋯⋯⋯⋯⋯⋯⋯⋯⋯⋯⋯⋯⋯ 265
3.2 简单包心菜叠层石（*Cryptozoon haplum* Liang, 1979）⋯⋯⋯ 273
3.3 加尔加诺锥叠层石（*Conophyton garganicum* Koroljuk, 1963）⋯⋯ 276
3.4 树桩圆柱叠层石（*Colonnella cormosa* Komar, 1964）⋯⋯⋯⋯ 283
3.5 喀什喀什叠层石（*Kussiella kussiensis*(Masliv)Krylov, 1963）⋯ 287
3.6 圆柱朱鲁莎叠层石（*Jurusania cylindrica* Krylov, 1963）⋯⋯ 290
3.7 地窖印卓尔叠层石（*Inzeria intia* Walter, 1972）⋯⋯⋯⋯⋯⋯ 295

2.2 DAWOKENG FORMATION ⋯⋯⋯⋯⋯⋯⋯⋯⋯⋯⋯⋯⋯⋯ 124
2.3 KUANGSHISHAN FORMATION ⋯⋯⋯⋯⋯⋯⋯⋯⋯⋯⋯ 144
2.4 TAIZI FORMATION ⋯⋯⋯⋯⋯⋯⋯⋯⋯⋯⋯⋯⋯⋯⋯⋯⋯ 184
2.5 YEMAHE FORMATION ⋯⋯⋯⋯⋯⋯⋯⋯⋯⋯⋯⋯⋯⋯⋯ 226
2.6 WENSHUIHE FORMATION ⋯⋯⋯⋯⋯⋯⋯⋯⋯⋯⋯⋯⋯ 233
2.7 SHICAOHE FORMATION ⋯⋯⋯⋯⋯⋯⋯⋯⋯⋯⋯⋯⋯⋯ 240
2.8 THE END OF STROMATOLITIC GREAT BARRIER REEFS IN SHENNONGJIA ⋯⋯⋯⋯⋯⋯⋯⋯⋯⋯⋯⋯⋯⋯⋯⋯⋯⋯ 253

3 MESOPROTEROZOIC STROMATOLITES FROM SHENNONGJIA ⋯⋯⋯⋯⋯⋯⋯⋯⋯⋯⋯⋯⋯⋯⋯⋯⋯⋯⋯ **263**
3.1 *Megadomia shennongjiaensis* Qian et al., 2017 ⋯⋯⋯⋯⋯ 265
3.2 *Cryptozoon haplum* Liang, 1979 ⋯⋯⋯⋯⋯⋯⋯⋯⋯⋯⋯ 273
3.3 *Conophyton garganicum* Koroljuk, 1963 ⋯⋯⋯⋯⋯⋯⋯ 277
3.4 *Colonnella cormosa* Komar, 1964 ⋯⋯⋯⋯⋯⋯⋯⋯⋯⋯ 283
3.5 *Kussiella kussiensis*(Masliv)Krylov, 1963 ⋯⋯⋯⋯⋯⋯⋯ 287
3.6 *Jurusania cylindrica* Krylov, 1963 ⋯⋯⋯⋯⋯⋯⋯⋯⋯⋯ 290
3.7 *Inzeria intia* Walter, 1972 ⋯⋯⋯⋯⋯⋯⋯⋯⋯⋯⋯⋯⋯⋯ 295

3.8 瘤通古斯叠层石（*Tungussia nodosa* Semikhatov, 1962）…………300
3.9 贝加尔贝加尔叠层石（*Baicalia baicalica* (Maslov) Krylov, 1963）
………………………………………………………………303
3.10 育卡贝加尔叠层石（*Baicalia unca* Semikhatov, 1962）…………306
3.11 奥姆泰尼奥姆泰尼叠层石（*Omachtenia omachtensis* Nuzhnov, 1967）………………………………………………………309
3.12 波层叠层石（*Stratifera undata* Komar, 1966）……………………312

结语 ……………………………………………………………316
致谢 ……………………………………………………………318
参考文献 ………………………………………………………319

3.8 *Tungussia nodosa* Semikhatov, 1962…………300
3.9 *Baicalia baicalica* (Maslov) Krylov, 1963 …………303
3.10 *Baicalia unca* Semikhatov, 1962…………306
3.11 *Omachtenia omachtensis* Nuzhnov, 1967 …………309
3.12 *Stratifera undata* Komar, 1966 …………312

CONCLUSIONS …………316
ACKNOWLEDGEMENTS …………318
REFRENCES …………319

0 引言
0 INTRODUCTION

神农架位于湖北省西北部，地处湖北省、重庆市交界，长江、汉水之间，是由大巴山脉东延余脉组成的中高山地貌。山峰多在海拔1 500 m以上，海拔2 500 m以上山峰有20多座，海拔3 000 m以上山峰有6座，最高峰神农顶海拔3 105.4 m，是华中地区最高峰。神农架山势雄伟，峡谷幽深，垂直高差达2 700 m，具有"一山分四季，十里不同天"的立体气候。

神农架处于大巴山常绿林生态区内，面积3 253 km²，是中国西部山地与东南部丘陵地的过渡区，也是中国东西南北多种植物区系成分的交汇区。拥有当今世界中纬度地区唯一保存完好的亚热带森林生态系统，森林覆盖率达96%，山区和湿地动植物区系成分丰富多彩，其中包括许多受保护的珍稀物种：珙桐（*Davidia involucrata*）、冷杉（*Abies fabri*）、铁坚杉（*Keteleeria davidiana*）、岩柏（*Selaginella moellendorfii*）、梭罗（*Reevesia pubescens*）、金丝猴（*Rhinopithecus roxellana*）、白化熊（*Ursus thibetanus*）、苏门羚（*Capricornis*

Shennongjia is located in the northwestern Hubei Province, between the Yangtze River and Han River, bordering Chongqing Municipality to the west. It is a landform of middle and high mountains considered to be the eastern (and the highest) section of the Daba Mountains. Most of mountains are higher than 1,500 m elevation, of which twenty odd peaks higher than 2,500 m, six tallest peaks are higher than 3,000 m. The highest peak is Shennongding peak, the first peak in Central China, is 3,105.4 m elevation. Majestic mountains, deep valleys, the vertical high amounts to 2,700 m, with a three-dimensional climate of "one mountain at four seasons, different weather within five kilometers".

Shennongjia lies within the Daba Mountains evergreen forests ecoregion, and occupies 3,253 km². It is located in the transition region of China's western mountains and southeastern hills, and is the intersection region of a variety of flora in the northern, southern, western and eastern China. It is the only well-preserved subtropical forest ecosystem in the world's middle latitudes. Forest coverage rate reaches 96%, various flora and fauna live in the district's mountains and wetlands, including many protected plant and animal species, such as *Davidia involucrata*, *Abies fabri*, *Keteleeria davidiana*, *Selaginella moellendorfii*, *Reevesia pubescens*, *sumatraensis*)、大鲵(*Andrias davidianus*)、白鹳(*Ciconia ciconia*)、白鹤(*Grus leucogeranus*)及金雕(*Aquila chrysaetos*)等。

神农架于1982年由湖北省批准,建立自然保护区。1986年7月由国务院批准,建立国家级森林和野生动物类型自然保护区。1990年12月17日,加入联合国教科文组织(UNESCO)人与生物圈计划世界生物圈保护区网。1995年,入选世界银行全球环境基金(GEF)资助的中国自然保护区管理项目示范保护区。2016年7月17日,第40届联合国教科文组织世界遗产委员会会议把中国湖北神农架列入世界遗产名录。

神农架除了丰富的动植物资源和秀美的景观外,还保存了非常珍贵的10多亿年前远古海洋微生物形成的生物礁遗迹。地质学家通过对这些遗迹不断深入的研究,逐步揭示地球的演化过程、生命的兴衰演替,为更好地保护我们唯一的家园——地球,提供科学的启迪。

Rhinopithecus roxellana, *Ursus thibetanus*, *Capricornis sumatraensis*, *Andrias davidianus*, *Ciconia ciconia*, *Grus leucogeranus* and *Aquila chrysaetos* etc.

Shennongjia Nature Reserve was established by Hubei Provincial Government in 1982, and was upgraded to be a state-level nature reserve in the category of forest and wild animals by the State Council of the People's Republic of China in July 1986. It was listed as a member of the Word's Protection Circle of Human and Animals by United Nations Educational, Scientific and Cultural Organization (UNESCO) on December 17, 1990, and as one of the demonstration projects of the China Nature Reserve Management Program by Global Environment Facility (GEF) in 1995. The 40th session of the World Heritage Committee decided to put China's Hubei Shennongjia on the prestigious World Heritage List on July 17, 2016.

In addition to various living flora and fauna as well as beautiful landscapes, there are very precious remains of ancient bioherms and biostromas built by marine microbes as far back as more than 1,000 million years ago in Shennongjia. By studing continuously these remains, geologists will gradually reveal the evolution process of the Earth and its biosphere to provide scientific enlightenment for protecting our only home — the Earth.

图0.1 （摄影:钱迈平）　　　　　　　　Fig. 0.1　（Photograph by Qian Maiping）

"华中屋脊"神农架,峰峦叠翠,云蒸霞蔚,气象万千,景色壮丽。

Shennongjia, known as "The Roof of the Central China", where marvelous scenery of magnificent mountains, virgin forest and deep canyons is very beautiful.

图 0.2　（摄影：钱迈平）　　　　　　　　　　Fig. 0.2　（Photograph by Qian Maiping）

　　在神农架的崇山峻岭中，时常可见10多亿年前远古海洋沉积形成的各种碳酸盐岩，白色的，灰色的，青色的，黄色的，红色的，黑色的，应有尽有。

Various carbonate rocks distributed in high mountains and lofty hills in Shennongjia. These rocks with white, grey, azury, orange, red and black colors formed in primal ancient marine deposits more than 1,000 million years ago.

图0.3 （摄影：钱迈平）　　　　　　　　Fig. 0.3　（Photograph by Qian Maiping）

仔细观察这些碳酸盐岩，还会发现其中一些呈现着各种神秘的精美纹层，或平缓，或曲折，或波澜起伏，或柱状排列，千姿百态，变幻莫测。这就是**叠层石**（stromatolites）的化石。

If you look at these carbonate rocks close-up, you will see some of the rocks appear mysterious, fine-layered structures exhibiting a variety of morphologies, including stratiform, conical, branching, domal, and columnar types. These are fossil **STROMATOLITES**.

1 叠层石概述
1 BRIEF INTRODUCTION OF STROMATOLITES

1.1 叠层石是什么？

叠层石是由分泌黏液的微生物，特别是蓝细菌（cyanobacteria），组成的微生物席（microbial mat）在周期性的生命活动和沉积作用期间，捕获、黏结和胶结沉积物颗粒，在浅水或潮湿环境形成的一层一层叠置增生的生物沉积构造。

1.1 WHAT ARE STROMATOLITES?

Stromatolites are layered bio-chemical accretionary structures formed in shallow water or moist habitats by the trapping, binding and cementation of sedimentary grains by microbial mats of microorganisms, especially cyanobacteria.

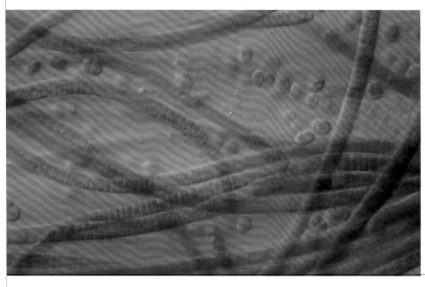

图1.1.1 （图片来源：galleryhip.com）

Fig. 1.1.1 （Photo credit: galleryhip.com）

蓝细菌是一种细胞中没有细胞核的原始微生物，和其他细菌一样进行分裂繁殖。这是它们在显微镜下的样子。

Cyanobacteria are a group of primordial microbiota without cell nuclei and reproduce by fission as other bacteria. This is what they look like under a microscope.

图1.1.2 （图片来源：luontoportti.com） Fig. 1.1.2 （Photo credit: luontoportti.com）

蓝细菌通过细胞中的光合色素，利用阳光、二氧化碳和水制造淀粉作为自己的营养，并释放氧气。其中，固氮蓝细菌还会通过细胞中的固氮酶，将大气中的氮还原成氮素化合物，并不断分泌氨基酸、多肽等含氮化合物，加上它们死亡后释放的大量氨态氮，都成为促进植物生长的肥料，大大改善了植物生长的环境。因此，蓝细菌的光合作用和固氮作用对地球早期充满二氧化碳和氮的缺氧大气圈改造起了巨大作用。

Most cyanobacteria are photoautotrophic organisms (some are also photoheterotrophic, which means they use light to generate ATP but they must obtain carbon in organic form) that fix CO_2 and release O_2. Another unique characteristic of some cyanobacteria is their ability to fix elemental (gaseous) nitrogen, and cause ammonia nitrogen released after their death and improve the environment for plant growth. Thus, cyanobacteria played a significant role in the transformation of the early Earth's anoxic atmosphere rich in carbon dioxide and nitrogen.

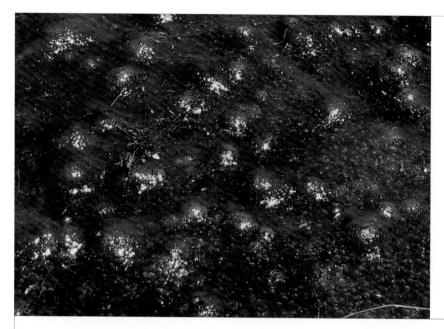

图 1.1.3
(图片来源:Jim Conrad)

Fig. 1.1.3 (Photo credit: Jim Conrad)

　　这是现代还在生长的叠层石,其顶面是微生物席。微生物席是由许许多多蓝细菌等微生物构成的多层状薄层,并由它们分泌的黏糊糊的多糖物质胶结在一起。

The top layer of a living stromatolite is a microbial mat consisted of multi-layered sheet of microorganisms and held together by slimy substances (polysaccharides) secreted by them.

图 1.1.4
(图片来源:sepmstrata.org)

Fig. 1.1.4 (Photo credit: sepmstrata.org)

　　微生物席下面,是它们一层层向上生长的同时所捕捉、黏结和胶结的一层层砂和矿物质。这种构成叠层石的层,地质学术语称为基本层(lamina)。

Layers of a stromatolite are formed by the trapping, binding and cementation of sedimentary grains by microbial mats of microorganisms, especially cyanobacteria. The layer of a stromatolite is known as lamina in geological terms.

图 1.1.5 （摄影：马雪） Fig. 1.1.5 （Photograph by Ma Xue）

如果在叠层石化石上切下一个纵切面的薄片，放在显微镜下观察，可以看到这些基本层是由一层暗色有机质降解物和一层亮色矿物结晶交替叠置而成的，这实际上记录了微生物席生命活动的周期性变化。

Under a microscope, the longitudinal section of the stromatolite shows alternation of bright sparry and dark micritic laminae represented the periodic growth rhythms of the microbial mat during its life activities.

1.2 叠层石化石是我们直接用肉眼就可看到的地球最古老的生命记录

叠层石早在地球诞生初期就已出现,可以毫不夸张地说,如果没有当初叠层石的大繁荣,就没有今天地球的蓝天白云和万物生长!

为什么这么说?这就要从宇宙的演化说起——

1.2 STROMATOLITE FOSSILS ARE THE EARTH'S OLDEST LIFE RECORD VISIBLE TO OUR NAKED EYES

Stromatolites can be traced back to the early Earth. It can be said without exaggeration that no prosperity of stromatolites in early Earth, no blue sky and white cloud as well as everything grows in Earth today!

How so? This begins with the development of the universe —

图 1.2.1 (图片来源: seeker.com)

Fig. 1.2.1 (Image credit: seeker.com)

约137亿年前,宇宙诞生于一次大爆炸!

宇宙大爆炸理论是目前科学界最著名,也是被广泛接受的宇宙演化模型,其依据来自长期的天文观测和计算研究。主要证据有:

(1)银河外天体有系统性谱线红移,红移与距离成正比。也就是说,它们正以与距离成正比的速度离开我们,距离越远,视向速度越大。用多普勒效应来解释,这显然是宇宙膨胀的反映,也暗示了宇宙曾经被压缩得很小。爱德温·哈勃(Edwin Hubble,1889—1953年)在1929年发现了这一现象,这就是天文学界称为的哈勃定律(Hubble's Law)。

(2) 如果宇宙大爆炸的确发生过,那么爆炸的最初时刻一定是非常非常热的,宇宙中应留有这种热的残余。1965年,美国射电天文学家阿尔诺·彭齐亚斯(Arno Penzias)和罗伯特·威尔逊(Robert Wilson)发现了一处2.725 K(-454.765℉,-270.425℃)的宇宙微波背景辐射(CMB)弥漫在可观测到的宇宙中,认为这就是科学家们一直在寻找的热残余。他俩因这一发现分享了1978年的诺贝尔物理学奖。

(3) 在可观测到的宇宙中,丰富的"轻元素"氢和氦随处可见,这也支持了大爆炸模型。

根据大爆炸模型的描述,约137亿年前,物质和能量浓缩在一个极小的奇点,温度极高,密度极大,瞬间发生快速膨胀。于是,包含着物质、能量、时间和空间的宇宙就诞生了,并且一直在不断地膨胀,温度则相应下降,从而形成了星系、恒星、行星乃至生命。

13,700±200 million years ago: estimated age of the universe.

How the universe developed into what it is today? One of the most famous and widely accepted models for the universe's development is the Big Bang Theory. This theory developed from observations and calculations of the structure of the universe as well as from theoretical considerations by astronomers and cosmologists. Evidences supported the Big Bang Theory:

(1) Galaxies appear to be moving away from us at speeds proportional to their distance. This is called "Hubble's Law", named after Edwin Hubble (1889—1953) who discovered this phenomenon in 1929. This observation supports the expansion of the universe and suggests that the universe was once compacted.

(2) If the universe was initially very, very hot as the Big Bang suggests, we should be able to find some remnant of this heat. In 1965, radio astronomers Arno Penzias and Robert Wilson discovered a 2.725 degree Kelvin (-454.765 degree Fahrenheit, -270.425 degree Celsius) Cosmic Microwave Background radiation (CMB) which pervades the observable universe. This is thought to be the remnant which scientists were looking for. Penzias and Wilson shared the 1978 Nobel Prize in Physics for their discovery.

(3) The abundance of the "light elements" hydrogen and helium found in the observable universe are thought to support the Big Bang model of origins.

The idea is that, since galaxies are moving away from each other now, then they used to be closer to each other. Extrapolating backwards in time, one can go back far enough to a time and place where all matter resided in a very small, dense region, called a singularity.

This state would have been all energy, as matter could not have existed as we know it in an energy field so intense. The Big Bang was the release of that energy, creating the current era of space time. The energy dispersed and, as things cooled down, matter began to form. It's the expanding universe that is the evidence for the Big Bang.

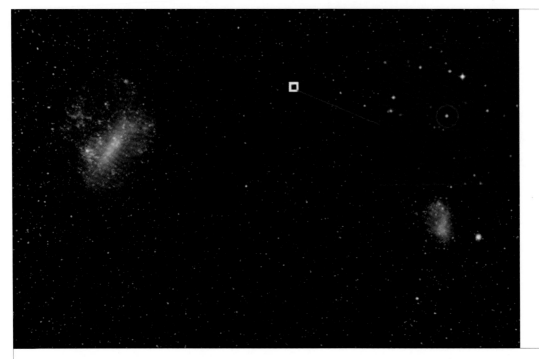

图1.2.2 （图片来源：Keller et al., 2014）　　　Fig. 1.2.2 （Photo credit: Keller et al., 2014）

约136亿年前，已知宇宙中最早的恒星形成。

2014年2月9日，澳大利亚天文学家斯特凡·科勒尔(Stefan Keller)的研究团队宣布，他们用赛丁泉天文台的"天图绘制者"(SkyMapper)望远镜，在进行天图绘制者南半球天空调查项目(SkyMapper Southern Sky Survey，简称SMSS)期间，发现了已知宇宙中最古老的恒星，命名为SMSS J031300.36-670839.3，简称SM0313。这颗星形成于大爆炸后仅1亿年左右，由更古老的恒星死亡时的超新星爆发释放出的气体云聚集形成，距离地球约6 000光年，位于水蛇座。

About 13,600 million years ago: the oldest known star was formed in the universe.

On February 9, 2014, astronomer Stefan Keller and his team announced their discovery of one of the oldest stars known in the Milky Way's galactic halo with the SkyMapper Telescope at Siding Spring Observatory in Australia. Found as part of the SkyMapper Southern Sky Survey, the star was designated SMSS J031300.36-670839.3, shortened to SMSS0313. The star was formed from the remnants of one of the universe's very first supernovae, just more or less 100 million years after the Big Bang, and located about 6,000 light-years away from Earth, in Constellation Hydrus.

图 1.2.3 （图片来源：eso.org/public）　　Fig. 1.2.3 （Image credit: eso.org/public）

约131亿年前，银河系形成。

多数天文学家认为，银河系和所有其他星系一样，诞生于大爆炸后产生的气体坍缩。起初，气体和暗物质粒子一起被引力束缚形成一个近似球形的巨大质量团块，其中的气体原子因其耗散性而迅速失去能量，向内坍缩。由于角动量总是守恒的，因此坍缩到质量团块内部的气体旋转速度加快，形成一个转动的气体盘，这就是银河系的胚胎。很快气体因辐射能量而冷却下来，团块凝聚成更小尺度，最终触发了第一代恒星的形成，银河系从此被点亮。

About 13,100 million years ago: the Milky Way was formed.

Most astronomers believe that the space was filled with hydrogen and helium after Big Bang, while some areas were slightly denser than others. The denser areas slowed the expansion slightly, allowing gas to accumulate in small protogalactic clouds. In these clouds, gravity caused the gas and dust to collapse and form stars. These stars burned out quickly and became globular clusters, but gravity continued to collapse the clouds. As the clouds collapsed, they formed rotating disks. The rotating disks attracted more gas and dust with gravity and formed galactic disks. Inside the galactic disk, new stars formed. What remained on the outskirts of the original cloud were globular clusters and the halo composed of gas, dust and dark matter.

图 1.2.4 （图片来源：NASA/JPL-Caltech/T. Pyle） Fig. 1.2.4 （Image credit: NASA/JPL-Caltech/T. Pyle）

约45.66亿年前，太阳系形成。

多数天文学家相信，太阳系和所有其他恒星系一样，诞生于星际云团的收缩聚集。他们推断，在银河系某处，一大片星际云团在引力作用下收缩聚集成一团。位于中心很小区域内的气体，在引力挤压下形成超高密度和温度的球体，这就是太阳的胚胎。

引力作用持续而强烈，气体和尘埃被不断吸入并相互加压，产生越来越多的热量。太阳胚胎变得更小、更亮、更热，内核开始产生核聚变反应。巨大的能量向四周喷出，形成强大的能量风，吹离了四周尚未来得及吸入的气体和尘埃，太阳就这样形成了。

太阳形成后，周围大量的星际气体、岩石和冰块，在太阳引力作用下围绕太阳公转。在太空零重力状态下，它们不会四散悬浮，而是会在引力作用下聚成一团，形成最初的行星。

那时，有数以百计的新生行星围绕太阳运行，拥挤而混乱，相互撞击频繁，有的彼此融合成更大的行星，有的粉身碎骨，有的被更大的行星俘获而成为它的卫星，最终形成太阳系。

About 4,566 million years ago: the Solar System was formed.

The formation of the Solar System began with the gravitational collapse of a small part of a giant molecular cloud. Most of the collapsing mass collected in the center, forming the Sun, while the rest flattened into a protoplanetary disk out of which the planets, moons, asteroids, and other small Solar System bodies formed.

The strong and persistent gravity pulled the gas and dust together, forming a solar nebula. The cloud began to spin as it collapsed, and eventually the cloud grew hotter and denser in the center, where the central mass became so hot and dense that it eventually initiated nuclear fusion in its core. As the cloud continued to fall in, the center eventually got so hot that it became a star, the Sun, and blew most of the gas and dust of the new solar system with a strong stellar wind.

A disk of gas and dust surrounding it that was hot in the center but cool at the edges. As the disk got thinner and thinner, particles began to stick together and form clumps. Some clumps got bigger, as particles and small clumps stuck to them, eventually forming the first planets.

Hundreds of newly formed planets surrounding in the Sun were so crowded and chaotic that they impacted each other frequently. Some of them got bigger, and other ones were smashed to pieces, or became moons of bigger planets, as they collided each other, eventually forming the Solar System.

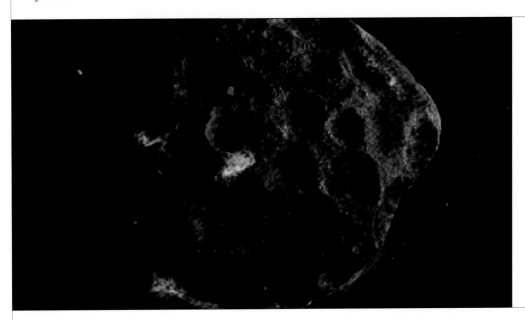

图 1.2.5 （图片来源：BBC）   Fig. 1.2.5 （Image credit: BBC）

约45.4亿年前,地球诞生。

天文学研究显示,围绕初生的太阳旋转的原始星云,经不断引力集聚、碰撞和挤压,逐渐形成了内部灼热的行星雏形,其中包括地球胚胎。

About 4,540 million years ago: the Earth was formed.

From astronomers' observations and estimates, the materials inside the primordial nebula revolving around the primary Sun drew together, bound by the force of gravity, into larger particles. The solar wind swept away lighter elements, such as hydrogen and helium, from the closer regions, leaving only heavy, rocky materials to create smaller terrestrial planets like Earth.

图1.2.6 (图片来源:Bullet-Magnet)　　　Fig. 1.2.6 (Image credit: Bullet-Magnet)

随后,地壳开始形成。

地球诞生后,其中的放射性物质发生衰变,使地球内部进一步升温。当温度升到铁的熔点时,大量融化的铁向地心沉降,并以热的方式释放重力能。大量的热使地球内部广泛融化并发生改变,逐步形成分层结构,其中地心是致密的铁核,熔点低的较轻物质则浮在表面,经冷却形成地壳。地球的磁场或许就是在这个时期形成的。

Subsequently, Earth's crust was forming.

Earth's rocky core formed first, with heavy elements colliding and binding together. Dense material sank to the center, while the lighter material created the crust. The planet's magnetic field probably formed around this time.

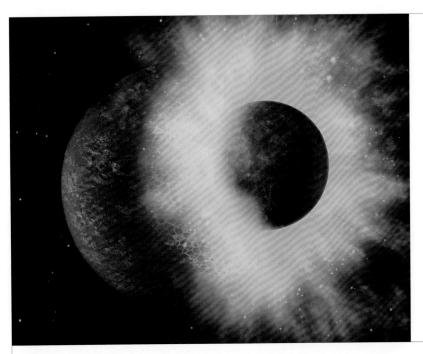

图 1.2.7　（图片来源：A. Passwaters / Rice University / NASA / JPL-Caltech）
Fig. 1.2.7　(Image credit: A. Passwaters / Rice University / NASA / JPL-Caltech)

约44亿年前，地球与一颗正在形成中的行星迎头相撞，产生了月球。

据美国"阿波罗"飞船采集的月球岩石样品，以及其他各项研究结果，多数科学家认为，地球形成1亿多年时，曾遭到一颗比地球小的、名叫忒伊亚（Theia）的行星撞击！飞溅起的熔融碎块，有一部分被地球的引力捕获进环绕地球运转的轨道，最终聚集形成月球。

新生的月球距离地球比现在近得多，在那时的地球上仰望天空，可看到一个巨大的月亮，可能比现在看到的大10—20倍。因此，当时月球对地球的引力比现在大得多，导致早期地球表面的岩浆及随后形成的海洋都有超强的潮汐作用。

About 4,400 million years ago: a head-on collision between Earth and a forming planet gave birth to Moon.

A Mars sized planet once orbited the Sun not far away from Earth. This early planet named Theia collided with Earth. The collision resulted in Theia being partially absorbed into earth, but a significant amount of debris from both Theia and Earth were sprayed around our planet. Gravity pulled the debris into orbit around Earth and as the fragments collided, they began to quickly coalesce together to form Moon.

The newly formed Moon was orbiting so much closer to the Earth that from the Earth it could see the Moon 10 to 20 times as huge as today's, causing a gravitational force much more strong than today's and resulting super strong tides in magma oceans, and then water oceans, on the it surface of the early Earth.

图 1.2.8　（图片来源：Ron Miller via International space art network）
Fig. 1.2.8　（Image credit: Ron Miller via International space art network）

　　地质学家将距今 45.4 亿－40 亿年地球的初生时期称为冥古宙（Hadean Eon），这个名字取自希腊神话的地狱之神冥王（Hades），以此形容当时地球表面环境像地狱一般——遍地熔融的岩浆，超高强度的宇宙辐射，还有陨石和小行星等太阳系其他天体的狂轰滥炸……

　　The Hadean is a geologic eon of the Earth, and it began with the formation of the Earth about 4,540 million years ago and ended, as defined by the International Commission on Stratigraphy, 4,000 million years ago. The name "Hadean" comes from Hades, the ancient Greek god of the underworld, in reference to the hellish conditions on Earth at the time: the planet had just formed and was still very hot due to high volcanism, a partially molten surface and ultra-high energy cosmic rays, as well as frequent collisions with other Solar System bodies.

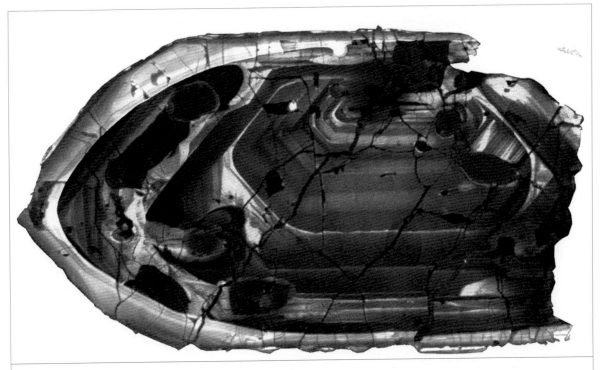

图1.2.9 （图片来源：John, 2014）   Fig. 1.2.9 (Photo credit: John, 2014)

    这是约44亿前、已知最古老的地壳碎片。

    在西澳大利亚杰克山(Jack Hills)冥古宙岩石中所发现的微小锆石晶体碎屑，经放射性同位素测定和原子探针断层摄影研究相互印证，确定其年龄为约44亿年。也就是说，地球形成1亿多年后，表面已冷却，地壳已形成。

    研究还发现，这颗锆石碎屑来自花岗闪长岩或石英闪长岩，因这两种岩石都富含水，所以推测当时的地表已存在液态水。

    About 4,400 million years ago: The oldest-known fragment of Earth's crust.

    This is a small fragment of a mineral called zircon extracted from a Hadean rock outcrop in western Australia's Jack Hills region crystallized more than 100 million years after the Earth formed.

    Trace elements in the oldest zircons from Australia's Jack Hills range suggest they came from water-rich, granite-like rocks such as granodiorite or tonalite, suppose where temperatures were low enough for liquid water, water not long after the planet's crust congealed from a sea of molten rock.

图 1.2.10 （图片来源：nature.com） Fig. 1.2.10 (Image credit: nature.com)

地球水圈和大气圈形成。

那时聚合在地球内部及彗星撞击带来的水，受热汽化上升，再冷却成云降雨。大雨可能连续不断地下了几百万年，其中夹杂着无数的闪电，不断催化着岩石中的氮、氢等元素，形成氨基这样的组成低级生命所必需的有机分子。随着雨水不间断侵入，地表逐渐冷却，氨基酸等大分子形成，原始的海洋也随之诞生。那时气温仍很高，甚至海水是沸腾的。

地球有相当大的质量，所产生的引力足以捕获一些气体，组成最初的大气圈。随着地球大气圈的形成，大量的陨石和小行星在进入大气圈后，因其超高速的运动与大气摩擦产生高温而烧蚀掉，而后天体轰炸逐渐减弱。当然，原始的大气圈是缺氧的，主要成分是混有氢气和水蒸气的高密度二氧化碳。臭氧层还没有形成，地球表面的宇宙辐射依然很强烈！

Earth's hydrosphere and atmosphere formed.

When Earth formed 4,540 million years ago from a hot mix of gases and solids, it had almost no atmosphere. The surface was molten. As Earth cooled, an atmosphere formed mainly from gases spewed from volcanoes. It included hydrogen sulfide, methane, and 10 to 200 times as much carbon dioxide as today's atmosphere. Gravity captured some of the gases releasing from volcanoes that made up this planet's early atmosphere. Hot volcanic gases built up the original atmosphere, then cooled over about half a billion years, Earth's surface cooled and solidified enough for water to collect on it. This is believed to have contributed half of the water that built up Earth's earliest oceans. This process further cooled the planet's surface and created some more stable land areas. The ancient atmosphere was toxic and filled with poisonous gases, and the oceans were filled with organic molecules, such as amino acids, nitrogenous bases, and hydrocarbons. They were also filled with heavy metals, specifically large amounts of Iron.

This planet contains proportionately more surface water than comparable bodies in the inner solar system. Outgassing of water from the Earth's interior is not sufficient to explain the quantity of water. One hypothesis that has gained popularity among scientists is that the early Earth was subjected to a period of bombardment by comets and water-rich asteroids. Much of the water on the surface today is thought to have originated from the outer parts of the solar system, such as from objects that arrived from beyond Neptune.

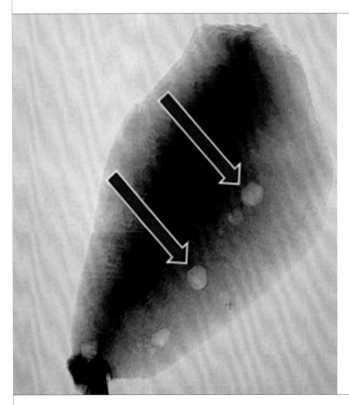

图1.2.11 （图片来源:Elizabeth et al., 2015）

Fig. 1.2.11 (Photo credit: Elizabeth et al., 2015)

在杰克山冥古宙岩石中,又发现了一粒微小的锆石,包含有几个石墨斑点,测定其碳12/碳13的比值,居然与光合作用产物的一致,而测定这颗锆石的年龄是41亿年!因石墨是在锆石结晶时包裹进去的,应该比锆石更古老,表明那时可能已经有生物在进行光合作用.

The zircon from a Hadean rock outcrop in Jack Hills contains inclusions of 4,100 million-year-old graphite. The graphite is older than the zircon containing it, and has a characteristic signature — a specific ratio of carbon-12 to carbon-13 — that maybe indicates the presence of photosynthetic life.

图1.2.12 (图片来源:MarioProtIV)　　Fig. 1.2.12 (Image credit: MarioProtIV)

约40亿年前,地球历史进入了太古宙(Archean Eon)。

虽然,布满大大小小陨石和小行星撞击坑的地球表面仍有局部的岩浆在涌动,但大部分已被原始的海洋覆盖,并渐渐形成了最初的大陆。

About 4,000 million years ago: the Archean Eon began.

The Earth's crust had cooled enough to allow the formation of continents, and this planet's surface had be mostly covered by water, though large and small impact craters could be found everywhere, as well as partial lava flows.

图 1.2.13 （图片来源：arcadiastreet.com）   Fig. 1.2.13 （Image credit: arcadiastreet.com）

  地质学家将距今 40 亿－25 亿年的地质时期称为太古宙（Archean Eon），以地壳形成和地球上出现最古老的岩石为起点。名字取自古希腊字"开始"的意思，并根据地球演化特点进一步划分为始太古代（距今 40 亿－36 亿年）、古太古代（距今 36 亿－32 亿年）、中太古代（距今 32 亿－28 亿年）和新太古代（距今 28 亿－25 亿年）。那时的地壳已冷却到足以形成大陆和海洋，而且随着大气圈屏障的完善，天体的狂轰滥炸也大大缓解，生命开始形成。

  The Archean Eon is a geologic time recognized by the International Commission on Stratigraphy, beginning with the formation of Earth's crust and the oldest Earth rocks about 4,000 million years ago and continuing until 2,500 million years ago. Archean comes from the ancient Greek Αρχή, meaning "beginning". The geologic time scale is divided into four classes of measured time: Eoarchean (4,000 to 3,600 million years ago), Paleoarchean (3,600 to 3,200 million years ago), Mesoarchean (3,200 to 2,800 million years ago) and Neoarchean (2,800 to 2,500 million years ago). During the Archean, the Earth's crust had cooled enough to allow the formation of continents and oceans. Meteor bombing was consumedly mitigated as the atmosphere was continuously perfected, and life started to form.

图1.2.14 （图片来源：Allen, et.al., 2016）

Fig. 1.2.14 （Photo credit: Allen, et. al., 2016）

  始太古代，约37亿年前，在今天格林兰西南端沿海的伊苏阿岛，构成叠层石的基本单元——微生物席已经出现，这是目前可用肉眼直接看到的最古老生命的记录。

A probable 3,700 million-year-old Eoarchean fossilized microbial mat, an origin of stromatolite, in Isua island off the southwestern tip of Greenland is the oldest life record visible to our naked eyes.

图1.2.15 （图片来源：aca.unsw.edu.au）

Fig. 1.2.15 （Photo credit: aca.unsw.edu.au）

  此后，叠层石陆续在世界各地生长发育。

  古太古代，约34.8亿年前，在西澳大利亚的皮尔巴拉（Pilbara）地区，德莱塞组（Dresser Formation）淡水热泉沉积硅质岩中保存的层柱状叠层石化石，是已知最古老的陆地生命记录。

Subsequently, stromatolites colonized gradually around the world.

About 3,480 million years ago: the oldest known evidence of life on land.

The fossil stromatolites from Paleoarchean freshwater hot spring deposits of Dresser Formation in the Pilbara, western Australia, were the earliest signs of life on land.

图1.2.16 （图片来源：Djokic, et al., 2017） Fig. 1.2.16 （Photo credit: Djokic, et al., 2017）

将其磨制成薄片，在显微镜下可见这些34.8亿年前陆地淡水热泉周期性叠层生长的微生物席化石，以及保存在其中的气泡构造。

A microscopic image of geyserite textures and spherical bubbles of the stromatolites from Paleoarchean Dresser Formation in the Pilbara, western Australia. This shows that surface hot spring deposits once existed there 3,480 million years ago.

图1.2.17 （图片来源：bgc-jena.mpg.de）

Fig. 1.2.17 (Photo credit: bgc-jena.mpg.de)

古太古代，约34.3亿年前，西澳大利亚的皮尔巴拉地区，斯特雷利湖组（Strelley Pool Formation）岩石中的层锥状叠层石化石。

About 3,430 million years old stromatolites from Paleoarchean Strelley Pool Formation in Pilbara region, West Australia.

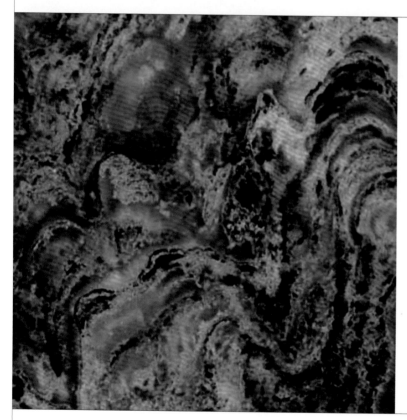

图1.2.18 （图片来源：crystalworldsales.com）

Fig. 1.2.18 (Photo credit: crystalworldsales.com)

打磨抛光后看，这些34.3亿年前的微生物席因光合作用而向上趋光生长造就的纹层更显美妙绝伦！

A polished section of the 3,430 million years old stromatolite shows beautiful laminae formed by photosynthetic microbiota mats expressed upward phototaxis.

图1.2.19 （图片来源：Som, et al., 2016） Fig. 1.2.19 (Photo credit: Som, et al., 2016)

新太古代，约27亿年前，西澳大利亚比斯利河（Beasley River）地区的穹状叠层石化石。测定同地点同时期熔岩流进海里快速冷却时捕获的气泡，结果显示当时的大气压不足今天的一半，说明那时的大气圈还很薄，宇宙射线对地表的照射还非常强，但并不影响叠层石的茁壮生长。

About 2,700 million years ago: huge Neoarchean domed stromatolites from Beasley River area in West Australia show evidence of single-celled, photosynthetic life on the shore of a large lake. A research uses bubbles trapped in the rocks suggested that air at that time exerted at the most half the pressure of today's atmosphere.

图1.2.20 （图片来源：all-geo.org） Fig. 1.2.20 (Photo credit: all-geo.org)

约25亿年前，叠层石大堡礁。

发现于南非北部古元古代沉积的德兰斯瓦超群（Transvaal Supergroup）玛尔玛尼白云岩（Malmani Dolomite）地层，由大型－巨型的穹状叠层石构成，显示此时叠层石已在一些地区形成规模宏大的生物礁。

About 2,500 million years ago: a stromatolite great barrier reef.

The great barrier reef built by giant domed stromatolites in Paleoproterozoic Malmani Dolomite of the Transvaal Supergroup in northern South Africa.

1.3 叠层石与有氧大气圈及多细胞生物的兴起

地球最初的大气圈的组分主要以宇宙中最丰富的氢气、氦气为主,很像木星的大气圈和太阳的组分。随着地球逐步冷却,其内部的剧烈岩浆活动释放出的气体主要以还原性气体为主,如水蒸气、二氧化碳及甲烷等,没有氧气。随着建造叠层石的蓝细菌等微生物的出现和繁盛,它们进行光合作用所释放的氧气在地球大气圈中日益积累,同时它们又大量消耗二氧化碳,使大气圈的组分发生了重大改变,为此后多细胞生物的大发展准备了必不可少的先决条件。

1.3 STROMATOLITES, OXYGENATED ATMOSPHERE AND THE RISE OF MULTICELLULAR ORGANISMS

Earth's first atmosphere was most likely comprised of hydrogen and helium that two most abundant gases found in the universe, and was very similar to the atmosphere of Jupiter and to the composition of the sun. Through the process of outgassing by volcanoes as cooling the Earth, the outpouring of gases from it interior, many other gases injected into the atmosphere. These include water vapor, carbon dioxide and ammonia, without oxygen. Light from the Sun broke down the ammonia molecules released by volcanos, releasing nitrogen into the atmosphere. Then, with the rise of photosynthetic microbial mats formed stromatolites, oxygen being released and accumulated, meanwhile carbon dioxide being consumed, the composition of the Earth's atmosphere had changed significantly and prepared an essential prerequisite for the rise of multicellular organisms.

图 1.3.1　（图片来源：ebay.de）　　　　Fig. 1.3.1　（Photo credit: ebay.de）

约 24 亿年前，"大氧化事件"。

25 亿年前至 5.42 亿年前，地球历史进入元古宙（Proterozoic Eon）。元古宙的命名源自古希腊字，意思是"较早期的生命"。分为古元古代（距今 25 亿－16 亿年）、中元古代（距今 16 亿－10 亿年）和新元古代（距今 10 亿－5.42 亿年前）。这一时期首次出现了一些稳定的大陆并开始相互拼接，还留存有大量的细菌和藻类化石的记录，其最重大的演化事件之一是有氧大气圈的形成。

古－中元古代期间，世界许多地方都大量沉积着一种铁矿，地质学上称为条带状铁建造（Banded Iron Formation，缩写 BIF）。其特点是交替沉积的氧化铁薄层，厚度为毫米至厘米级，既有磁铁矿（四氧化三铁 Fe_3O_4）也有赤铁矿（三氧化二铁 Fe_2O_3），其间夹泥岩和燧石（二氧化硅 SiO_2），经常呈红色。今天世界上我们已看不到这种类型的岩石形成，这就意味着形成它们的古海洋条件与今天的完全不同。

目前大多数科学家的解释是：早期地球的海洋是酸性的，溶解了大量来自海底火山提供的铁。早期地球大气圈是缺氧的，尽管这些铁在水中早已饱和，却没有析出沉淀。但随着建造叠层石的蓝细菌等微生物的兴起，它们进行光合作用所释放的氧气一旦遇到水中溶解的铁，就会形成不溶于水的氧化铁,并沉淀下来。因光合作用受日夜或季节或其他条件的周期性变化控制,氧的供给和铁的沉积也随之周期性变化,于是形成了 BIF。

About 2,400 million years ago: Great Oxygenation Event.

The period of Earth's history that began 2,500 million years ago and ended 542 million years ago is known as the Proterozoic, which is subdivided into three eras: the Paleoproterozoic (2,500 to 1,600 million years ago), Mesoproterozoic (1,600 to 1,000 million years ago), and Neoproterozoic (1,000 to 542 million years ago). The name Proterozoic comes from Greek and means "earlier life". During the Proterozoic, stable continents first appeared and began to accrete. Also coming from this time are the first abundant bacteria and algae fossils. One of the well-identified events of this eon was the transition to an oxygenated atmosphere.

During Paleo-Mesoproterozoic, a type of sedimentary iron deposit named Banded Iron Formations (BIF) had been found throughout the world. A typical BIF consists of repeated, thin layers (a few millimeters to a few centimeters in thickness) of silver to black iron oxides, either magnetite (Fe_3O_4) or hematite (Fe_2O_3), alternating with bands of iron-poor shales and cherts (SiO_2), often red in color, of similar thickness, and containing microbands (sub-millimeter) of iron oxides. We do not see any rocks of this type forming in the world today, and this suggests that conditions in the ancient oceans where they formed were quite different from today's.

It is assumed that initially the Earth started with vast amounts of iron dissolved in the world's acidic seas. As photosynthetic organisms generated oxygen, the available iron in the Earth's oceans precipitated out as iron oxides forming a thin layer on the ocean floor, which may have been anoxic mud (forming shale and chert). Each band is similar to a varve, to the extent that the banding is assumed to result from cyclic variations in available oxygen, then formed BIF.

图1.3.2 （图片来源：Graeme Churchard）　　　　Fig. 1.3.2 （Photo credit: Graeme Churchard）

虽然BIF早在30多亿年前的太古宙时期就已零星出现，那时叠层石是零星分布的，氧气的产量很低，但距今24亿—19亿年期间BIF却在世界范围大量涌现，这正是叠层石分布扩展、藻类植物兴起、氧气产量剧增的时期，这就是地球演化史上著名的"大氧化事件（Great Oxygenation Event）"。

BIF appeared widely 2,400 to 1,900 million years ago when stromatolites extending throughout the world and various algae rising, thus leading to a surge production of oxygen, although a little of BIF had found more than 3,000 million years ago with rare stromatolites. This was a result of famous Great Oxygenation Event.

图 1.3.3 （图片来源：microbiology.ubc.ca）　　　　Fig. 1.3.3　（Photo credit: microbiology.ubc.ca）

　　BIF 是世界上最重要的铁矿床，分布十分广泛，有不少大型、特大型矿床，如北美的苏必利尔湖铁矿、西澳大利亚的皮尔巴拉铁矿、中国的鞍山铁矿、俄罗斯及乌克兰的库尔斯克－克里沃罗格铁矿等。BIF 占世界铁总储量的 60% 以上，占富铁矿储量的 70% 之多。

　　BIF are the most important commercial source of iron ore, including some giant ore deposits such as the Lake Superior iron ores in North America, the Pilbara ores in West Australia, the Anshan iron ores in China and the Kursk-Krivoy Rog iron ores in Russia and Ukraine etc. BIF accounts for more than 60% of the world's total iron reserves, and for 70% of the rich iron ore reserves.

图1.3.4 （图片来源：El Albani A, et al., 2010）

Fig. 1.3.4 （Photo credit: El Albani A, et al., 2010）

这是约21亿年前，已知最古老的多细胞生物。

随着大气圈氧气的增加，以及抵御宇宙辐射的臭氧层的形成，这时的地球才真正成为拥有蓝天碧海的美丽星球，不再只有简单的显微镜才能看到的微小单细胞生物，复杂的毫米一厘米级多细胞生物应运而生。如2010年公布的非洲加蓬的弗朗斯维尔附近约21亿年沉积岩中发现的，数以百计的10—12 cm大小形状各异的多细胞软躯体生物化石。

About 2,100 million years ago: oldest known multicellular organisms.

The centimetre-sized structures from the black shales of the Palaeoproterozoic Francevillian B Formation in Gabon of Africa, are interpreted as highly organized and spatially discrete populations of colonial organisms. The structures are up to 12 cm in size and have characteristic shapes, with a simple but distinct ground pattern of flexible sheets and, usually, a permeating radial fabric. Geochemical analyses suggest that the sediments were deposited under an oxygenated water column. The growth patterns deduced from the fossil morphologies suggest that the organisms showed cell-to-cell signalling and coordinated responses, as is commonly associated with multicellular organization.

图 1.3.5 (图片来源:Forme Di Vita Aliena Su Marte Il Daily Mail E Le Immagini Picture)
Fig. 1.3.5 (Image credit: Forme Di Vita Aliena Su Marte Il Daily Mail E Le Immagini Picture)

距今18亿—9亿年,叠层石进入最繁荣时期。

叠层石经过10多亿年的演化发展,到中元古代时期,进入了最繁荣时期。其间约12.5亿年前,叠层石无论在规模数量上还是在形态的多样性上,都达到繁荣的顶峰,几乎占领了世界各地的的浅海、浅滩、潮坪、盐坪、潟湖及热泉等水域,并在许多地方构成绵延数千米至数十千米的大堡礁,是地球大气圈氧气的重要供应者。

About 1,800 million years ago to 900 million years ago: the most prosperous period of the history of stromatolites.

After more than a billion years of evolution, stromatolites had peaked in widely distributed and morphological diversity during Mesoproterozoic. They grown in shallow seas, beaches, tidal flats, sabkhas, lagoons and hot springs etc. all around the world, even formed great barrier reefs stretching dozens to hundreds of kilometers in some shallow waters. They were important suppliers of oxygen in the Earth's atmosphere.

图 1.3.6 （图片来源：Zhu et al., 2016）　　　　Fig. 1.3.6 （Photo credit: Zhu et al., 2016）

  这是约 15.6 亿年前，高于庄生物群。
  化石发现于中国河北燕山山脉南麓中元古代高于庄组海相沉积地层，以多细胞毫米—厘米级藻类为特征。藻类形态包括带状、舌状、楔形和长卵形等多种。最大的舌状化石长达 28.6 cm，宽近 8 cm；最大的带状化石长度可达 30 cm 以上，宽达 4.5 cm；有的标本可见明显的底部固着器官，表明它们是固着在海底竖直生长的。由此可见，当时藻类植物已经非常繁盛，它们也是地球大气圈氧气的重要供应者。

  About 1,560 million years ago: Gaoyuzhuang Biota.
  The centimetre-sized linear, cuneate, oblong and tongue-shaped carbonaceous compressions from grey-black shales of Mesoproterozoic Gaoyuzhuang Formation in Yanshan area, North China, that have statistically regular linear to lanceolate shapes, of which the largest tongue-shaped up to 28.6 cm long and nearly 8 cm wide, while the largest linear shape more than 30 cm long and up to 4.5 cm wide suggesting that the Gaoyuzhuang fossils record benthic multicellular eukaryotes of unprecedentedly large size. Syngenetic fragments showing closely packed about 10 μm cells arranged in a thick sheet further reinforce the interpretation. Comparisons with living thalloid organisms suggest that these organisms were photosynthetic, which were also the main supplieis of oxygen in the Earth's atmosphere.

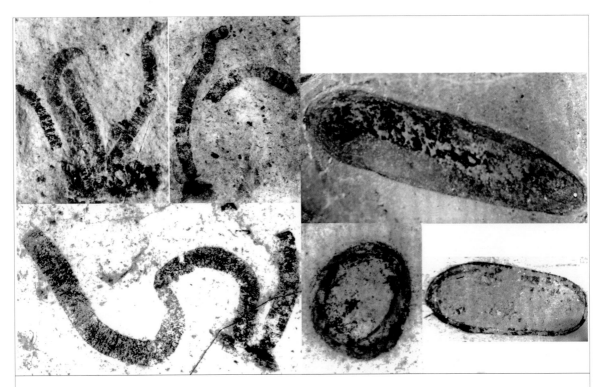

图 1.3.7 （图片来源：钱迈平等，2009）　　　　Fig. 1.3.7 （Photo credit: Qian et al., 2009）

这是 8 亿多年前，徐淮生物群。

化石发现于中国江苏、安徽北部及山东南部新元古代海相沉积地层，以多细胞毫米－厘米级的球状、卵状、囊状、带状浮游及固着生长藻类为特征。

More than 800 million years ago: Xuhuai Biota.

Between millimetre and centimetre-sized spherical, axiolitic, cystic and linear-shaped carbonaceous compressions from grey-black shales of Neoproterozoic formations in Xuhuai area, southern margin of North China, are interpreted as multicellular macroscopic plankton and attached algae according to their morphological characteristics.

图 1.3.8 （图片来源：NASA/SPL）　　　　Fig. 1.3.8 （Image credit: NASA/SPL）

距今7.5亿—6.35亿年，发生了"雪球地球事件"。

距今7.2亿—6.35亿年，新元古代晚期，地球至少经历了3次（或许4次）大冰期，地史学上称这段时期为"成冰纪（Cryogenian）"。地层中保存的特征性冰碛沉积显示当时地球曾经历了最严酷的冰期，其冰盖多次扩展和退缩，甚至可能到达赤道海洋。其中以距今7.5亿—7亿年的斯图特冰期（Sturtian Glaciation）及6.35亿年前结束的马林诺冰期（Marinoan Glaciation）规模最大，几乎波及全球，这就是地球历史上著名的"雪球地球事件（The Snowball Earth Event）"。可以想像，那时的地球成了一个冰雪包裹的星球，在茫茫宇宙中疾驰达数百万至数千万年！在这样的大冰期中，大批生物群灭亡。除了耐寒生物群外，其他残存的生物群（包括形成叠层石的微生物群）被分割在很局限的避难所内，如火山地域的热液喷口及热泉附近，继续生息繁衍，相互隔绝数百万至数千万年，各自独立演化。

About 750 to 635 million years ago: the Snowball Earth.

About 720 to 635 million years ago, there were three (or possibly four) significant ice ages during the late Neoproterozoic, that is Cryogenian period. Characteristic glacial deposits indicate that Earth suffered the most severe ice ages in its history during this period. Glaciers extended and contracted in a series of rhythmic pulses, possibly reaching as far as the equator. At least two major worldwide glaciations of which, the Sturtian glaciation persisted from 750 million years ago to 700 million years ago, and the Marinoan glaciation which ended approximately 635 million years ago. This is generally called "Snowball Earth". Imagine the earth hurtling through space like a cosmic snowball for millions to dozens of millions of years. Apart from psychrophilic (cold-loving) organisms, a large number of other organic communities were killed by the Snowball Earth event. The survivors (including microbiotas formed stromatolites) clinging precariously to hydrothermal vents or hot springs within particular volcanic fields might maintain a high degree of genetic isolation for millions to dozens of millions of years, and evolved respectively.

图1.3.9 （图片来源:Hoffman and Schrag, 2000）　　Fig. 1.3.9 （Photo credit: Hoffman and Schrag, 2000）

雪球地球事件是严寒到酷热急剧转换的环境巨变，其完整的沉积地层序列自下而上是："复理石建造"（巨厚的砂岩和泥岩）或火山岩，冰碛岩及含铁或锰沉积岩，"盖帽"碳酸盐岩，以及含磷沉积岩。

雪球地球事件的成因至今仍未完全研究清楚，通常的解释是：

距今12.5亿－8.2亿年发生的一系列大规模造山运动，如北美的格林威尔运动（The Grenville Orogeny）、北欧的挪威运动（The Sveconorwegian Orogeny）及中国的晋宁运动等，导致许多地区的地壳升降剧烈。而当时地球的陆地都处于赤道和低纬度地区，气候炎热多雨，抬升地区的岩石侵蚀作用强烈，产生的巨量的陆源碎屑被河流带入海洋，沉积下巨厚的砂岩和泥岩。同时，强烈的侵蚀作用大量消耗着大气中的温室气体二氧化碳，导致气温迅速下降，地球两极冰盖扩大。

到约7.2亿年前，开始进入成冰纪。在约7.5亿年前，当时地球上一个

The Snowball Earth Event was a striking freeze-fry climate reversal, which had been recorded in the Cryogenian rocks. The sedimentary succession from bottom to top including: thick layers of sandstone and mudstone, or volcanic rocks, deposits of iron-rich and manganese-rich mixed with the glacial beris, "cap" carbonates, as well as deposits of phosphate-rich.

The cause of the Snowball Earth has not yet been completely made clear. An explanation for it accepted by most of geologists as follow:

About 1,250 million years ago to 820 million years ago, a series of orogenies, such as the Grenville Orogeny in North America, the Sveconorwegian Orogeny in North Europe and the Jinning Orogeny in China, would have leaded to large structural deform of the Earth's crust. The continents had clustered together near the equator in the Meso-Neoproterozoic, where the climate was hot and rainy. As a result of orogenies, some areas would have been uplifting rapidly and being intensely eroded, while other areas would have been subsiding sharply and being deposited by thick layers of sandstone and mudstone. Intense erosion in continents near the equator would have excessively consumed atmospheric carbon dioxide — a greenhous gas and have rapidly reduced global temperatures.

The breakup of the supercontinent Rodinia roughly 750 million years ago, caused large amounts of river water to rush into the oceans. This fundamentally changed the ocean chemistry to cause gigantic build-up of

巨大的陆地——罗迪尼亚超大陆（Rodinia），在地壳构造运动中分裂解体，导致大量的河水涌入海洋，从根本上冲淡了海水的盐度，从而更容易造成冰层的形成和扩张。超大陆的裂解还导致了一些地区持续发生大规模的火山喷发，喷出的大量二氧化硫到达对流层顶，甚至进入平流层，便很难通过降雨回到地面，因而长期滞留高空，阻挡了太阳的辐射，也会导致地球气温下降，两极冰盖扩大。

随着地球冰盖从极地向低纬度扩张，辐射到地球上的阳光被冰雪大量反射，使气温进一步下降，又进一步导致冰盖扩大。一旦冰盖越过地球30°纬度，且面积超过地球表面的50%，冰雪对太阳辐射的反射就开始加速加剧，产生失控的反射反馈效应（runaway albedo feedback），一发不可收地致使地球气温加速下降，直至赤道海洋冻结。此时地球平均气温下降到约-50℃，要知道今天地球的平均气温约为15℃。可想而知，那时的

ice. As another result of the breakup of the supercontinent, volcanoes would have continuously erupted through sulfur-rich sediments, which would have been pushed into the atmosphere during eruption as sulfur dioxide. When sulfur dioxide gets into the upper layers of the atmosphere, and past the tropopause, the boundary separating the troposphere and stratosphere, it is most effective at blocking solar radiation. If it reaches this height, it is less likely to be brought back down to earth in precipitation or mixed with other particles, extending its presence in the atmosphere from about a week to about a year. If this eruption were to go on for a decade or so, the volcanoes may have added enough sulfur dioxide aerosols into the atmosphere to destabilize the climate and rapidly cooling the planet.

The Solar radiation interacts with the Earth's surface and atmosphere to control climate. The more radiation the planet reflects, the cooler the temperature. White snow and ice reflects the most solar energy and has a high albedo, and they cool the atmosphere and thus stabilize their own existence. This phenomenon, called the ice-albedo feedback, helps polar ice sheets to grow. With temperatures reducing, the ice sheets would have been expanding from hight latitudes to lower latitudes. When the ice sheets formed at latitudes lower than around 30 degrees north or south of the equator, the planet's albedo began to rise at a faster rate because direct sunlight was striking a larger surface area of ice per degree of latitude. The feedback became so strong in this simulation that surface temperatures plummeted and the entire planet froze

地球是多么地寒冷！而且地质记录显示,这样的冰封可能持续上千万年！

同时,因冰雪覆盖,岩石的化学风化作用减弱,对大气中二氧化碳的消耗大大减少。陆地火山不断喷出的二氧化碳,在大气中逐渐积累达到很高的浓度,直至达到现代大气二氧化碳含量的350倍！地球开始解冻,冰盖从低纬度向两极退缩。同样,冰盖一旦退过地球30°纬度,即启动与失控的反射反馈相反的效应,使地球气温加速升高,直至地球完全解冻,进入酷热环境,全球平均气温约50℃。

由于地球的板块构造运动一直都没停止,沿着相互撕裂的洋中脊以及相互抵触的岛弧—海沟,分布着许多海底火山,它们喷涌的岩浆不断改变着海水的化学性质。上千万年的冰封使大气中的氧不能溶入冻结的海洋,全球水域长期严重缺氧。随岩浆喷出的铁、锰等元素,不能形成氧化物沉淀,很容易溶解进水里,逐渐积累到高浓度状态。随着洋面迅速解

over. Once the Earth had entered a deep freeze, the high albedo of its icy veneer would have driven the average temperature on the surface of Earth to so low, about −50 ℃, which it seemed there would have been no means of escape.

Although it seems the planet might never wake from its cryogenic slumber, subaerial volcanoes slowly manufacture an escape from the chill: carbon dioxide, that is one of several gases emitted from volcanoes. Normally this endless supply of carbon is offset by the erosion of silicate rocks: the chemical breakdown of the rocks converts carbon dioxide to bicarbonate, which is washed to the oceans. There bicarbonate combines with calcium and magnesium ions to produce carbonate sediments. During a global glaciation, shifting tectonic plates would continue to build volcanoes and to supply the atmosphere with carbon dioxide. At the same time, the liquid water needed to erode rocks and bury the carbon would be trapped in ice. With nowhere to go, carbon dioxide would collect to incredibly high levels—roughly 350 times the today concentration of carbon dioxide—high enough to heat the planet and end the global freeze. Once melting begins, low-albedo seawater replaces high-albedo snow and ice, and the runaway freeze is reversed.

Millions to dozens of millions of years of ice cover would deprive the oceans of oxygen, so that dissolved iron and manganese expelled from the seafloor hot springs could accumulate in the water. Once a carbon dioxide-induced greenhouse effect began melting the ice, oxygen would mix with the

冻,大气中的氧气溶入海水,与这些高浓度铁、锰等元素发生反应,形成大量氧化物沉淀,在许多地方富集成矿。

在酷热的气候及特有的海水化学性质等一系列条件作用下,浅海形成了那个时期特有的"盖帽"碳酸盐岩,直接覆盖在解冻沉积形成的冰碛岩上。

雪球地球事件期间,幸存的生物群落会分散避难于世界各地可维持生存的零星区域,如热泉附近等地。当两个起初相同的群落被彼此分隔孤立超过几百万年,可能会通过某种程度的基因突变,产生新的物种。

雪球地球事件结束后,在距今6亿－5.8亿年,通过剧烈转换的严寒－酷热气候的多次筛选,幸存下来的生物类群随着气候环境趋向改善,数量开始回升,群落开始扩散并彼此融合,并迅速分异增殖,种类和数量空前增加。它们在当时特定环境下的生命活动,形成了大量磷氧化物,在许多地方富集成矿。

---

seawater again and force the iron and manganese to precipitate out with the debris once carried by the sea ice and glaciers.

The rapid termination would have resulted in a warming of the Snowball Earth to extreme greenhouse or hothouse conditions, upward to almost 50 ℃. The transfer of atmospheric carbon dioxide to the ocean would result in the rapid precipitation of calcium carbonate in warm surface waters, producing the "cap" carbonate rocks observed globally.

During glaciations, hot-spring communities widely separated geographically on the icy surface of the globe would accumulate genetic diversity over millions of years. When two groups that start off the same are isolated from each other long enough under different conditions, chances are that at some point the extent of genetic mutation will produce a new species. Repopulations occurring after each glaciation would come about under unusual and rapidly changing selective pressures quite different from those preceding the glaciation; such conditions would also favor the emergence of new life-forms.

A series of global freeze-fry events would have imposed an environmental filter on the evolution of life. Survivors of this calamity began to diversify and proliferate rapidly as the end of the hothouse period in the last Snowball Earth Event, with the climatic amelioration, and formed deposits of phosphate-rich under specified conditions about 600 to 580 million years ago.

图 1.3.10 （图片来源：jxdkj.gov.cn）

Fig. 1.3.10 (Photo credit: jxdkj.gov.cn)

"雪球地球"解冻时期是重要的成矿时期。

因各地区的化学环境不同，有的地区铁富集形成铁矿，如中国赣湘桂地区的"新余式铁矿"；有的地区锰富集形成锰矿，如中国湘鄂川黔地区的大塘坡组锰矿层。

The Snowball Earth thawing promoted ore deposits forming.

According to different geochemical environments, iron and manganese ore deposits formed, such as Xinyu Type Iron Deposits in Jiangxi, Hunan and Guangxi provinces, and manganese ore deposits of Datangpo Formation in Hunan, Hubei, Sichuan and Guizhou provinces, South China.

图 1.3.11 （图片来源：钱迈平等，2012）

Fig. 1.3.11 (Photo credit: Qian et al., 2012)

随后，生物群逐渐复苏和繁盛，产生出大量的磷，也在一些地区富集成矿，如中国贵州瓮安陡山沱组、江西广丰皮园村组磷矿层等。这些在雪球地球解冻后形成的沉积矿床，规模相当大，往往断续延伸上千千米，而且品位较高，是重要的矿产资源。

Redemption of life following the last Snowball Earth Event formed substantial deposits of phosphate-rich, such as the phosphate ore deposits of Doushantuo Formation in Weng'an, Guizhou Province and Piyuancun Formation in Guangfeng, Jiangxi Province, China etc. These ore deposits formed after the Snowball Earth thawing are commonly high-grade and significant mineral resources, extending for thousands of kilometers.

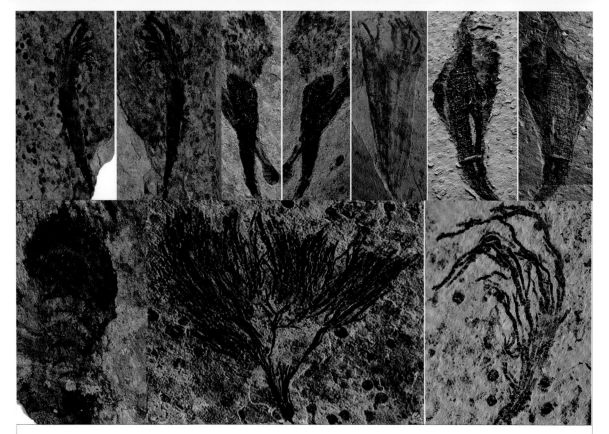

图1.3.12　（图片来源：Yuan et al., 2011）　　Fig. 1.3.12　（Photo credit: Yuan et al., 2011）

这是距今6.35亿—5.8亿年，蓝田生物群。

随着雪球地球事件结束，地球历史进入新的历史时期，即埃迪卡拉纪（Ediacaran period），时限在距今6.35亿—5.41亿年。这一时期，地球生命迅速向厘米级以上的更大个体和更加多样化发展。发现于中国安徽省休宁县蓝田镇附近的蓝田生物群就是其早期的代表，包括各种扇状和丛状生长的海藻，以及具有触手和类似肠道特征的动物，个体大多达3—4 cm长，大的可达15 cm。

About 635—580 million years ago: the Lantian Biota.

About 635 million years ago the Ediacaran period began at the end of the Cryogenian Snowball Earth, and it is followed by the Cambrian, the first period of the Paleozoic, about 541 million years ago. The period is famous for the first larger-bodied fossils, which are probably the first recorded metazoans. The Lantian Biota found near Lantian Town of Xiuning County, Anhui Province, is preserved in black shales of the Lantian Formation, and yields some of complex centimetre-sized organisms, including fan-shaped seaweeds and possible animal fossils with tentacles and intestinal-like structures reminiscent of modern coelenterates and bilaterians. Most of individuals are 3—4 cm long, of which the largest one up to 15 cm.

约5.85亿年前,发生了埃迪卡拉辐射(也称为阿瓦隆爆发)。

埃迪卡拉纪,始于6.35亿年前的成冰纪结束,终于5.42亿年前的寒武纪开始,标志着原始生物时代的元古宙结束和较高等生物时代的显生宙开始。其间发生的一次重大的生物演化事件,就是埃迪卡拉辐射(Ediacaran Radiation),或称阿瓦隆爆发(Avalon Explosion)。

About 585 million years ago: Ediacaran Radiation (Avalon Explosion).

The Ediacaran Period, from the end of the Cryogenian Period 635 million years ago to the beginning of the Cambrian Period 542 million years ago, marks the end of the Proterozoic Eon and the beginning of the Phanerozoic Eon. A great evolutionary event taken place during this period is the Ediacaran Radiation, also called the Avalon Explosion.

图 1.3.13 (图片来源: Ryan Somma)　　Fig. 1.3.13 (Image credit: Ryan Somma)

在约5.85亿年前,各种各样已知最古老的多细胞复杂生物快速地出现在世界各地。它们的个体大小通常为厘米级,甚至达到米级,最常见的有管状、棕榈叶状、盘状、垫状或囊状,大多固着生长,也有漂浮、游泳或爬行的,繁盛一时,这就是埃迪卡拉生物群(Ediacaran Biota)。它们都是软躯体生物,没有明显的口、肛门和消化道等器官分化,形态构造与后来显生宙出现的生物门类完全不同,甚至到现在,古生物学家们仍难以确定它们之间的系统演化关系。

在约5.42亿年前,随着寒武纪的开始,埃迪卡拉生物群突然大量消失。也正是在这时,随着地球大气圈中氧气的增加,新的更加丰富多彩的动物类型爆发性出现,它们具有发达的口、肛门和消化道等器官分化,有的还长出牙齿、甲壳或外骨骼,这就是寒武纪大爆发(Cambrian Explosion)。寒武纪大爆发出现的动物门类延续至今,而埃迪卡拉生物群却随之绝灭。

About 585 million years ago, various the earliest known centimetre-sized or even metre-sized complex multicellular organisms had rapidly occupied worldwide, the most common types of which resemble enigmatic tubes, fronds, disks, quilted mattresses or immobile bags. They were soft-bodied creatures, without distinct differentiation of organs such as the mouth, anus and digestive tract etc. These are known as the Ediacara Biota. It bears little resemblance to modern lifeforms, and their relationship even with the immediately following lifeforms of the Phanerozoic Eon is rather difficult to interpret.

About 542 million years ago, the Ediacaran Biota largely disappeared contemporaneously with the begining of the Cambrian, rising of the atmospheric oxygen level, and the rapid increase in biodiversity known as the Cambrian Explosion, in the event all modern animal phyla appeared. The creatures from Cambrian Explosion have been provided with distinct differentiation of organs such as mouth, anus and digestive tract etc., and some of them with teeth, shells or exoskeletons. Instead, the Ediacaran Biota died out rapidly.

图1.3.14 （图片来源：nature.com）　　　Fig. 1.3.14 （Image credit: nature.com）

约5.15亿年前，发生了寒武纪大爆发。

寒武纪从约5.42亿年前的埃迪卡拉纪结束，到4.854亿年前的奥陶纪开始，历时5 550万年。其间发生了地球生命演化史上的一次重大事件——"寒武纪生物大爆发（Cambrian Explosion）"！现代各类动物的祖先几乎一起涌现出来，例如，中国云南省澄江县约5.15亿年前，以及加拿大的不列颠哥伦比亚省布尔吉斯隘口附近约5.08亿年前的海相沉积岩石里，都发现了许多保存精美的各种各样的动物化石，它们被分别称为"澄江动物群"和"布尔吉斯动物群（Burgess Fauna）"。

About 515 million years ago: Cambrian Explosion.

The Cambrian Period lasted 55.5 million years from the end of the preceding Ediacaran Period 542 million years ago to the beginning of the Ordovician Period 485.4 million years ago. A significant evolutionary event, the Cambrian Explosion, occured during this period. It is the relatively short period in which many complex life forms appeared on the planet. It shows the sudden coming into existence of practically all known animal phyla. The Cambrian stratum also shows that there was a sudden and massive destruction of life because of the preservation of large quantities of fossils, of which exquisite preservation of rarely preserved, non-mineralized soft tissue, for example, the Chengjiang Fauna from the Maotianshan Shales (about 515 million years ago) in Chengjiang County, Yunnan Province, China and the Burgess Fauna from the Burgess Shales (about 508 million years ago) in the Canadian Rockies of British Columbia, Canada.

1.4 现代叠层石

叠层石在经历了约12.5亿年前的繁荣顶峰后，无论在规模数量上还是在形态多样性上，都开始持续衰退，特别到寒武纪生命大爆发后，随着各门类多细胞动物的大批兴起和发展，叠层石的领地大为缩小。现代叠层石分布非常罕见，仅生长在世界少数区域，而且形态也很单调，远没有中－新元古代时期那么丰富多彩。

1.4 MODERN STROMATOLITES

Stromatolites peaked about 1,250 million years ago and subsequently declined in abundance and diversity. Especially after the Cambrian Explosion with most major animal phyla appeared and prospered, their colonies reduced greatly. Modern stromatolites are rare and found only at a few sites.

西澳大利亚鲨鱼湾（Shark Bay）哈默林湖（Hamelin Pool）现代不分叉柱状叠层石。

Living columnar stromatolites in Hamelin Pool, Shark Bay, West Australia.

图1.4.1 （图片来源：sharkbay.org）　　Fig. 1.4.1　（Photo credit: sharkbay.org）

图1.4.2 （图片来源：soton.ac.uk）

Fig. 1.4.2 (Photo credit: soton.ac.uk)

卡塔尔的乌姆—赛伊德（Umm Said）附近萨布哈（Sabukha）盐坪现代层状叠层石。

Living stratiform stromatolites at the margin of a small lagoon, northern part of Umm Said Sabkha, Qatar.

图1.4.3 （图片来源：sepmstrata.org）

Fig. 1.4.3 Photo credit: sepmstrata.org

巴哈马群岛埃克马苏岛（Exumas）沿海的现代叠层石生物礁。

Modern stromatolite bioherms in the Exumas, Bahamas.

图1.4.4 （图片来源:en.wikipedia.org） Fig. 1.4.4 （Photo credit: en.wikipedia.org）

加拿大不列颠哥伦比亚省亭子湖（Pavilion Lake）的现代柱状叠层石。
Living columnar stromatolite towers at Pavilion Lake, British Columbia, Canada.

图 1.4.5 （图片来源：astrobio.net）

Fig. 1.4.5 （Photo credit: astrobio.net）

南极洲乔伊斯湖（Lake Joyce）水下 22 m 深度的现代分叉柱状叠层石。

Living columnar stromatolites with well-developed branches from about 22 m deep in Lake Joyce, Antarctica.

图 1.4.6 （图片来源：lagunabacalarinstitute.com）

Fig. 1.4.6 （Photo credit: lagunabacalarinstitute.com）

墨西哥尤卡坦半岛南部巴卡拉尔湖（Laguna Bacalar）巨大的现代叠层石生物丘。

Living giant stromatolite bioherms in Laguna Bacalar in southern Yucatán Peninsula in Mexico.

图1.4.7 （图片来源：Farías et al., 2013）

Fig. 1.4.7 （Photo credit: Farías et al., 2013）

阿根廷安第斯山脉海拔3 570 m索孔帕（Socompa）火山湖的现代层状叠层石。

The stratiform stromatolites developing at 3,570 m above sea level in a high-altitude Volcanic Lake Socompa, Argentinean Andes.

图1.4.8 （图片来源：nps.gov）

Fig. 1.4.8 （Photo credit: nps.gov）

美国黄石国家公园（Yellowstone National Park）肖松尼湖间歇喷泉盆地（Shoshone Lake Geyser Basin）热泉周围的现代丘状叠层石。

Modern domal stromatolites around a hot spring in Shoshone Lake Geyser Basin, Yellowstone National Park.

图 1.4.9 （图片来源：en.wikipedia.org）　　　　Fig. 1.4.9　（Photo credit: en.wikipedia.org）

　　澳大利亚新南威尔斯珍诺兰洞穴群（Jenolan Caves）洞穴里的现代丘状叠层石，高度和宽度都在 1 m 左右。

"Crayback" stromatolite — Nettle Cave, Jenolan Caves, New South Wales, Australia. Size is about 1 m high and wide.

1.5　叠层石成矿作用

　　大量的事实证明,许多成矿作用与微生物形成的叠层石有密切关系。
　　一方面,一些微生物的生命活动有助于某些矿物的富集和沉淀;另一方面,由于叠层石间隙及孔隙非常发育,而且叠层石的层理是由微生物席和沉积碎屑层交替形成的,因此层状叠层石层理沿水平方向延伸可构成开放结构,而丘状和柱状叠层石层理也可在较大范围内构成半开放结构,这些都有利于含矿热液的灌入和矿化。

1.5　STROMATOLITE-HOSTED MINERALIZATIONS

　　Plenty of facts prove that many mineralizations are closely related to form of stromatolites.

　　Activities of some microbes in the mats of stromatolites contribute to the enrichment and precipitation of some minerals. There are high porosity, on the other hand, and alternating laminae of microbial mat and sedimentary debris in stromatolites, thus open structures extending levelly formed by stratiform stromatolites and half-open structures occurring large areas formed by domical or columnar stromatolites. These make easily for hydrothermal fluid entering and mineralization.

图 1.5.1 （图片来源：fotos.etrr.com.br/algal-stromatolite）
Fig. 1.5.1 （Photo credit: fotos.etrr.com.br/algal-stromatolite）

1.5.1 叠层石富集镁

巴哈马的安德罗斯岛（Andros Island）发育大量现代叠层石，研究发现建造这些叠层石的某些蓝细菌能富集镁。这些蓝细菌在死亡后，会释放镁元素形成高镁方解石，并很容易使之白云岩化。这一发现可能有助于解释为什么大量的化石叠层石都以白云岩形式出现。

1.5.1　Magnesium enrichment of stromatolites

The study on many modern stroamtolites at Andros Island in Bahama found that some cyanobacteria can enrich magnesium from sea water during their building up these structures and release magnesium to form high magnesian calcites easily dolomitized after their death. Maybe this discovery helps explain why most of fossil stromatolites are in dolomites.

图 1.5.2
(图片来源：cfk-fossilien.de)

Fig. 1.5.2
(Photo credit: cfk-fossilien.de)

1.5.2 叠层石富集铁

元古宙几个主要含铁建造常与叠层石生物礁有关。

北美明尼苏达东北部的比瓦比克含铁岩组（Biwabik Iron Formation），包含两个燧石质柱状叠层石地层；安达略的冈费林特含铁岩组（Gunflint Iron Formation），由互层的燧石叠层石、层状燧石、铁燧石、碧玉、灰岩和黑色页岩组成。

南非的开普（Cape）和特兰斯瓦尔（Transvaal）含铁岩组通常盖在厚层状叠层石生物礁白云岩之上。

中国燕山西段宣化－龙关一带的长城系地层分布多种铁质叠层石生物礁化石，构成了铁含量高达46%－56%的宣龙式铁矿。

1.5.2 Iron enrichment of stromatolites

Several major Proterozoic iron formations are associated with stromatolite bioherms.

For example in North America, the Biwabik Iron Formation of northeastern Minnesota contains two siliceous columnar stromatolite beds, and the Gunflint Iron Formation of Ontario consists of interbedded chert stromatolites, bedded cherts, iron cherts, jaspers, limestones and black shales.

In South Africa, the Cape and Transvaal iron formations occurred usually on thick-bedded stromatolite bioherm dolostones.

In China, the Xuanlong Type Iron Ore Deposite (iron contains up to 46%－56%) is composed of various iron stromatolite boiherms from Changcheng System in Xuanhua-Longguan region.

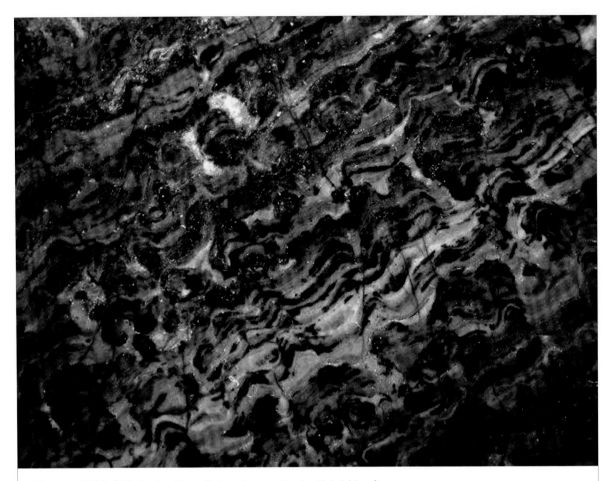

图 1.5.3 （图片来源：fossilmall.com/Science/stromatolite-fossils/michigan）
Fig. 1.5.3 （Photo credit: fossilmall.com/Science/stromatolite-fossils/michigan）

1.5.3 叠层石富集锰

例如，非洲刚果沙巴(Shaba)省古元古代地层中的层状锰矿床，主要由铁锰氧化带下的锰碳酸盐岩、锰榴石英和石墨片岩组成，其中锰碳酸盐岩的锰含量高达40%，并夹有多层叠层石生物礁化石。

美国密西根州北部巧克莱山(Chocolay Hills)古元古代巧克莱群(Chocolay Group)地层，其中锰碳酸盐岩也发育有菱锰矿叠层石生物礁化石。

1.5.3 Manganese enrichment of stromatolites

Such as the stratiform manganese deposits in Shaba, Congo in Africa are made up of Paleoproterozoic manganses carbonates, manganese garnet and graphite schists, of which, manganese content up to 40% in the manganese carbonates with stromatolite bioherms.

The rhodochrosite stromatolite bioherms in manganese carbonates from Paleoproterozoic Chocolay Group in Michigan, USA.

图 1.5.4 (图片来源:paleosearch.com/product/yelma-digitata-stromatolite-colonies)
Fig. 1.5.4 (Photo credit: paleosearch.com/product/yelma-digitata-stromatolite-colonies)

1.5.4 叠层石富集铅

西澳大利亚的纳伯鲁盆地(Nabberu Basin)古元古代约17亿年前的耶尔玛组(Yelma Formation)地层,丰富的含铅硫化矿床形成于微叠层石柱体之间。

1.5.4 Lead enrichment of stromatolites

Such as the rich lead sulfide deposits are formed between columns of Paleoproterozoic microstromatolites from Yelma Formation (about 1,700 million years old) in Nabberu Basin, West Australia.

图1.5.5 （图片来源：sjsresource.com.au/mineral-exploration-zambia-copper-belt）
Fig. 1.5.5 （Photo credit: sjsresource.com.au/mineral-exploration-zambia-copper-belt）

1.5.5 叠层石富集铜

刚果－赞比亚元古宙叠层石生物礁白云岩与沉积铜矿关系密切：叠层石生物礁本身铜含量贫乏，但与其毗邻的泥质板岩却富含铜矿，离生物礁较远的围岩含铜量明显减少，呈现叠层石生物礁 → 毗邻泥质板岩 → 较远围岩呈现斑铜矿 → 黄铜矿 → 黄铁矿化带的演替模式。

中国四川－云南交界地区的东川铜矿元古宙叠层石非常发育，铜的硫化物沿大型层状叠层石或大型柱状叠层石层理或层理间隙富集。

1.5.5 Copper enrichment of stromatolites

Proterozoic stromatolite bioherm dolostones are closely related to sedimentary copper ores in Congo-Zambia, where stromatolite bioherms themselves copper content are poor, but very rich in adjacent argillaceous slates, and the more far away from these bioherm dolostones are, the less copper in country rocks content. The stromatolite bioherm dolostones → adjacent argillaceous slates → far country rocks appeared bornite → chalcopyrite → pyritic zones

Proterozoic stromatolite bioherms are well developed in Dongchuan copper ore field in border areas of Sichuan and Yunnan, where copper sulfides enriched along laminae and their gaps of large stratiform and columnar stromatolites.

图1.5.6 （图片来源：ucl.ac.uk/earth-sciences） Fig. 1.5.6 (Photo credit: ucl.ac.uk/earth-sciences)

1.5.6 叠层石富集磷

印度乌代布尔（Udaipur）地区阿拉瓦利（Aravalli）元古宙沉积变质岩中，分布有磷块岩矿床。成矿的磷酸盐岩与层状和柱状叠层石（*Minjaria*、*Baicalia*及*Collenia*）紧密相关，叠层石中五氧化二磷（P_2O_5）含量在10%－37%，叠层石碎块形成角砾状磷块岩或硅质角砾岩。

在中国华南贵州开阳－息烽和湖北钟祥、保康及宜昌一带，埃迪卡拉纪陡山沱组磷矿床发育。其中一些叠层石生物礁本身就是磷矿床的组成部分，磷质在叠层石层理或柱体内更集中。

1.5.6 Phosphatised enrichment of stromatolites

The phosphorite deposits are closely related to stratiform and columnar stromatolites (*Minjaria, Baicalia* and *Collenia*) in metamorphic sediments of Proterozoic Aravalli Formation in Udaipur, India. The phosphorus pentoxide (P_2O_5) in these stromatolies content up to 10%－37%, and the stromatolite pieces formed phosphorite or siliceous breccias.

The phosphatic stromatolite bioherms from Ediacaran Doushantuo Formation themselves are parts of phosphorite deposits in Kaiyang-Xifeng of Guizhou, Zhongxiang, Baokang and Yichang of Hubei, where phosphorites become more enriched in laminates or columns of stromatolites.

图 1.5.7 （图片来源：wikipedia.org/wiki/Witwatersrand_Basin）

Fig. 1.5.7 （Photo credit: wikipedia.org/wiki/Witwatersrand_Basin）

1.5.7 叠层石富集金-铀

南非的威特沃特斯兰德（Witwatersrand，南非荷兰语：白水山脊）地区产金量占世界金产量相当的大比例。在卡尔顿维尔（Carletonville，以"巩固"金矿田的一位资深采矿总监盖·卡尔顿·琼斯的名字命名）金矿田布莱沃尤特齐什奇（Blyvooruitzicht，南非荷兰语：幸福前景）金矿，其金-铀矿沉积在太古代的主砾岩组（年龄约为29亿年）黑碳领导者段的碳质叠层石纹层里，是古微生物席降解后的产物。

1.5.7 Gold and uranium enrichment of stromatolites

The Witwatersrand (meaning the "ridge of white waters" in Afrikaans) area of South Africa produces a significant percentage of the world's gold. In Blyvooruitzicht (meaning the "happy prospect" in Afrikaans) Gold Mine of Carletonville (named after the long-serving mining director of Consolidated Gold Fields, Guy Carleton Jones) Gold field, native gold and radioactive uraninite enriched in intervals of stromatolites in blackened hydrocarbon-rich rocks of the Carbon Leader Member of the Archean Main Conglomerate Formation (about 2,900 million years old). This is the product by biodegradation of ancient microbial mats.

1.6 叠层石研究历史和现状

人们对叠层石的认识经历了一个逐步深入了解的漫长过程。

早在18世纪，一些科学家们就注意到，有些沉积岩石中存在着具有显著特征的向上隆起弯曲的多层纹层构造体，这就是所谓的叠层石。叠层石究竟是生物形成的，还是非生物沉淀的？起初科学家们还难以确定。

1.6 HISTORY AND CURRENT STATUS OF STUDY ON STROMATOLITES

Our gradual understanding of stromatolites have experienced a long process.

As early as the 18th century, some scientists and observers had noticed remarkable curved structures with multi-layered interiors in some sedimentary rocks. These are known as STROMATOLITES. Are stromatolites biogenic or abiogenic precipitates? Scientists could not affirm about this at that time.

图1.6.1 （图片来源：Michael C. Rygel/Wikimedia Commons）
Fig. 1.6.1 （Photo credit: Michael C. Rygel/Wikimedia Commons）

1825年，斯迪尔（J. H. Steel）首次描述了美国纽约州萨拉托加泉（Saratoga Springs）附近上寒武统霍伊特灰岩（Hoyt Limestone）中的纹层钙质体，认为可能是非生物沉积的"钙质结核构造"。1883年，霍尔（J. Hall）重新研究了它们，认为可能是简单的原始动物如层孔虫的骨架化石，将其命名为 Cryptozoon，意思是"隐秘动物石"。而从19世纪后期到20世纪初，科学家们对这种沉积构造的成因一直争论不休，有的认为它们是由类似海绵的动物所形成的，将其命名为 Spongiostroma，意思是"海绵叠层"；有的认为是由原生动物所形成的，将其命名为 Eozoon，意思是"原生动物石"；还有的认为是由类似珊瑚的动物所形成的，等等。

Upper Cambrian stromatolites in the Hoyt Limestone exposed at Lester Park, near Saratoga Springs, New York, were first described as abiogenic "calcareous concretions" by J. H. Steel in 1825. These were later named *Cryptozoon* by J. Hall who regarded them as the skeletons of simple animals such as stromatoporoid in 1883. Scientists debated keenly for many years on these during late 19th century and early 20th century. They thought these as sponge-like concretions that named *Spongiostroma*; or as protozoan products, named *Eozoon*; and or as coral-like structures etc.

最先揭示叠层石本质的是德国地质学家考科夫斯基（E. L. Kalkowsky）。

1908年他在发表的论文中,将德国北部哈尔茨山脉附近,下三叠统彩斑砂岩组（Buntsandstein Formation）湖相鲕粒岩地层中的柱状和丘状纹层构造称为叠层岩（Stromatolith）,即具有纹层的岩石,**叠层石（Stromatolite）**一词就是由叠层岩引申而来的。他还指出其纹层由纤维状丝体构成,呈不规则扇形排列,具有放射

图 1.6.2 （图片来源:dicciomed.eusal.es）
Fig. 1.6.2 （Photo credit: dicciomed.eusal.es）

1914年美国古生物学家及地质学家瓦尔科特（C. D. Walcott）将蒙大拿州大贝尔特山的中元古代硅质叠层石,以及纽约州萨拉托加泉附近的寒武纪叠层石,与纽约州湖泊里现代蓝细菌沉积作用形成的淡水钙华相比较,认为这些叠层石是沉积钙质的蓝细菌生命活动的产物,首次将叠层石与蓝细菌联系起来。许多后来的研究也确认了这个结论,瓦尔科特的观点得到普遍接受。

图 1.6.3 （图片来源:usbr.gov/history）
Fig. 1.6.3 （Photo credit: usbr.gov/history）

1930年3月英国地质学家布兰克（M. Black）研究巴哈马群岛安德罗斯岛海洋沉积期间,首次发现海相潮间带也存在叠层石,并于1933年发表了他的研究成果。此后,越来越多的滨海—浅海相叠层石（多数是化石叠层石）被发现,证明咸水环境也是叠层石的重要分布区域。

图 1.6.4 （图片来源:trinitycollegechapel.com）
Fig. 1.6.4 （Photo credit: trinitycollegechapel.com）

生长趋势；许多叠层石在生长形态上类似珊瑚和海绵。他认为这些穹形纹层的叠加、分叉柱体的生长是生物寻求光照和食物的反映，这是由一种能够导致碳酸钙沉淀的"类似植物的低等生物"形成的化石。

E. L. Kalkowsky, a German geologist, is the first to reveal the nature of the structures.

In 1908, he introduced the term Stromatolith—layered stone, that came to be called **Stromatolites**—to describe columns and domes of well layered carbonate within beds of Early Triassic Buntsandstein Formation's lacustrine oolite that occur near the Harz Mountains of northern Germany, and suggested that stromatolites were formed by "simply organized plant-like organisms".

C. D. Walcott, an American paleontologist and geologist, compared Mesoproterozoic stromatolites from Big Belt Mountains in Montana and late Cambrian stromatolites from Saratoga Springs in New York State with present-day freshwater tufas in lakes in New York State in 1914. He argued that Proterozoic and Cambrian stromatolites were "deposited through the agency of algae similar in type and activity to the (Cyanophyceae) Blue-green Algae". This was also the first link between stromatolites and cyanobacteria. His connection was confirmed by many subsequent researchers and became widely accepted.

M. Black, a British geologist, found firstly marine stromatolites at the margins of lakes and tidal creeks on Andros Island in March 1930 when he studied the marine deposits in the Bahamas. Since his work published in 1933, more and more coastal shallow marine stromatolites, especially fossil examples, have been discovered, proved that saltwater environment is also their important colonies.

图 1.6.5 （图片来源：uux.cn/viewnews） Fig. 1.6.5 （Image credit: uux.cn/viewnews）

　　叠层石是一种生物沉积构造，并非生物本身。单个叠层石并不代表单一生物。叠层石之所以引起许多科学家的浓厚兴趣，是因为它们不仅是地球上已知的肉眼可直接看见的最古老化石，而且也是我们深入窥视地球久远历史的独特窗口。由于不同的研究者因思考问题的角度不同，对叠层石形态变化规律的认识产生了很大分歧，出现了两大学派，即环境学派和古生物地层学派。

　　Stromatolites are biosedimentary structures rather than organisms. An individual stromatolite does not represent an individual organism. Many scientists have been particularly interested in stromatolites because they are not only the oldest fossils visible directly to our naked eyes, but our singular visual portal into deep time on Earth. There are very large differences on understanding of the morphologic changes of stromatolites due to different perpectives of researchers. Thus, there are the environmental school and the biostratigraphic school.

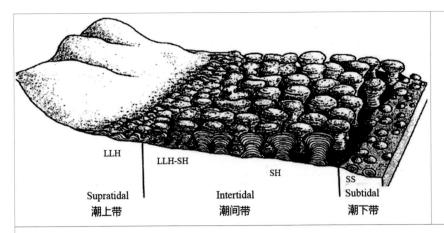

图 1.6.6 （图片来源：Selley, 1976）

Fig. 1.6.6 (Image credit: Selley, 1976)

1.6.1 环境学派

其主要观点由澳大利亚地质学家洛根（B. W. Logan）等人于1964年提出，曾受到沉积学者的广泛关注。他们根据分布非常局限的现代叠层石的研究，认为叠层石形态特征由沉积环境决定，与生物演化无关，所以不能用来确定其在地层上的先后顺序。

他们将叠层石分为3种基本类型：LLH型（侧向连接半球状）、SH型（分离的垂直叠加半球状）和SS型（分离的球状）。并认为LLH形成于有障壁潮坪，SH形成于时常露出水面的潮坪前岬，SS形成于动荡的潮坪较下部。

但环境学派至今仍无法解释化石叠层石，特别是寒武纪以前的叠层石，在形态多样性上远远超过现代叠层石的事实，也无法解释化石叠层石的某些形态组合只出现在寒武纪以前的某些地质时期，而且他们的叠层石分类方法也无法用来表述化石叠层石丰富多彩的形态类型，因此难以推广。

1.6.1 The environmental school

It was mainly proposed by Australian geologist B. W. Logan et al. in 1964, and had been widely noted by sedimentologists. Through study of present-day stromatolites, they believe that the morphological characteristics of stromatolites are determined by sedimentary environment and not related to biological evolution, so it can't be used to determine the order of their strata.

They classified stromatolites into three main types: LLH (laterally linked hemispheroids), SH (discrete, vertically stacked hemispheroids) and SS (discrete spheroids—either as randomly stacked hemispheroids or concentrically arranged spheroids). They considered that ancient environments may be interpreted by recognition of fossil stromatolite forms. For example, protected intertidal mud flats may be inferred by the presence of type-LLH stromatolites. Exposed, intertidal mud flats are inferred by the presence of type-SH structures. Low intertidal areas are inferred by the presence of type-SS structures.

However, they still can't explain why the diversity of fossil stromatolites, especially Precambrian examples, is far more than present-day's, so their classification can't adequately cover fossil examples.

图 1.6.7
(摄影:钱迈平)

Fig. 1.6.7 (Photograph by Qian Maiping)

1.6.2 古生物地层学派

以苏联地质学家克雷洛夫(I. N. Krylov)为代表,先后得到美国、澳大利亚、法国、印度和中国一些学者的继承和发展,他们主要致力于探索叠层石的形态学分类及地层分布规律。他们发现,某些叠层石组合在时代分布上是寒武纪以前的某一段时期专有的。

他们认为化石叠层石实质上是微生物遗迹化石,因此采用古生物学惯用的拉丁文双名法对叠层石进行命名分类,通常以叠层石形态或首次发现地点命名,基本上解决了对化石叠层石复杂形态变化的表述问题。他们按叠层石形态特征将其分为柱叠层石、非柱叠层石和微叠层石3个亚类。柱叠层石再进一步分为分叉柱状和不分叉柱状,非柱叠层石也进一步分为层状、丘状和球状,此外还有柱状和非柱状的混合类型。

1.6.2 The biostratigraphic school

The Soviet Union geologist I. N. Krylov is the representative for this school that it has been inherited and developed by some scholars from the United States of America, Australia, France, India and China. They focus on exploring morphological classification and stratigraphic distribution of fossil stromatolites, and have found some stromatolite assemblages only occurred within a limited range of the Precambian strata.

They considered fossil stromatolites in fact are trace fossils of microbial life that can be classified and named in the binomial nomenclature used in paleontology. They usually name a stromatolite with its morphological characteristics or first discovered site, which expressed well the complex morphologic changes of the stromatolite. They classified stromatolites into three main types: columnar stromatolites, non-columnar stromatolites and microstromatolites, and further divided columnar into branching and non-branching stromatolites, non-columnar into stratiform, domical and spherical stromatolites, in addition combinative types of columnar and non-columnar stromatolites.

2 神农架地质公园主要景区的中元古代叠层石大堡礁群遗迹

2 REMAINS OF THE MESOPROTEROZOIC STROMATOLITE GREAT BARRIER REEFS IN MAIN TOURIST SPOTS OF SHENNONGJIA UNESCO GLOBAL GEOPARK

中元古代是叠层石最繁荣兴旺的时期,全世界的浅水区域几乎到处都有它们形成的大量生物礁。在许多地方甚至出现绵延数十甚至上百千米的叠层石大堡礁群,当时的神农架地区也是其中之一。

The Mesoproterozoic was the most prosperous period of the history of stromatolites, and there were large numbers of bioherms made up of stromatolites in shallow waters around the world, some stromatolite bioherm groups as great barrier reefs in many places stretched for tens, even hundreds, of kilometers, such as in the Shennongjia.

图 2.0.1 （摄影：钱迈平） Fig. 2.0.1 （Photograph by Qian Maiping）

神农架在地理上位于渝鄂交界处的鄂西山区，毗邻湖北省竹山、房县、保康、兴山、巴东及重庆市巫山县边区。地理坐标为北纬31°15′—32°00′、东经109°56′—111°00′。属秦岭山系大巴山脉东段，山脉大致东西走向。地势西南高，东北低。峰高谷深，沟壑纵横，一般高差300—1 200 m，最大高差2 700 m，为侵蚀地貌及峡谷地形。

Shennongjia is located in the mountainous region in the northwestern part of Hubei Province, neighbouring Zhushan, Fangxian, Baokang, Xingshan, Badong counties of Hubei Province and Wushan County of Chongqing Municipality. Geographical coordinates: Latitude 31°15′—32°00′ N, Longitude 109°56′—111°00′ E. It is the eastern section of the Daba Mountains of the Qinling Mountain Range, and trend in general west-northwest to east-southeast direction, and is erosion landforms and canyon topography, the vertical high of mountains-valleys amounts in general to 300—1,200 m, the highest to 2,700 m.

图 2.0.2 （摄影：钱迈平）　　　　　　　　　　Fig. 2.0.2　（Photograph by Qian Maiping）

神农架中元古代叠层石大堡礁群，被很好地保存在10多亿年前沉积的浅海碳酸盐台地相地层中。目前，地质学家将这套地层称为神农架群，是一套以白云岩为主的岩石组合，其次还有砂岩、粉砂岩、砾岩、火山岩及铁矿石等，地层厚达12 680 m，有多次基性－中性火山岩浆活动记录。

In Shennongjia, the Mesoproterozoic great barrier reefs made up of stromatolite bioherm groups in an ancient shallow-marine carbonate platform were well preserved in sedimentary sequences more than 1 billion years. This sedimentary succession, known as Shennongjia Group by geologists, is mainly dolostones, in addition, there are sandstones, siltstones, conglomerates, volcanic rocks and iron ore layers etc., and is up to 12,680 m thick, with multiple basaltic-neutral magmatism records.

图 2.0.3 （摄影：钱迈平） Fig. 2.0.3 （Photograph by Qian Maiping）

神农架地质公园主要景区公路沿线出露的神农架群地层，自下而上包括：乱石沟组、大窝坑组、矿石山组、台子组、野马河组、温水河组和石槽河组等。

The Shennongjia Group sequences exposed along highways in main tourist spots of Shennongjia UNESCO Global Geopark in an ascending order: Luanshigou, Dawokeng, Kuangshishan, Taizi, Yemahe, Wenshuihe and Shicaohe formations etc.

2.1 乱石沟组

沿神农谷－神农顶的公路沿线：

乱石沟组厚 1 137 m，以各种白云岩为主，发育多层叠层石生物礁白云岩。

沉积构造以水平层理为主，具条带状构造、波状层理、重力滑动层理、波痕及冲刷面，同生角砾、碎屑、鲕粒、泥裂、岩盐假晶。

2.1 LUANSHIGOU FORMATION

Along Shennonggu-Shennongding highway:

Luanshigou Formation is 1,137 m thick in the section and consisted basically of various dolostones including several stromatolite bioherm-biostrome dolostone beds.

The sedimentary structures in this formation are dominated by banded horizontal bedding, along with current bedding, cross bedding, gravity sliding structures, ripple marks, erodsion surfaces, breccias, detritals, oolites, mud cracks and halite pseudomorphs.

图 2.1.1
(摄影:钱迈平)

Fig. 2.1.1
(Photograph by Qian Maiping)

乱石沟组下部灰黑色厚层－块状叠层石生物礁白云岩。白云岩是一种高镁碳酸盐沉积岩,含高比率的白云石矿物(碳酸镁钙),通常是热带海洋碳酸盐沉积物在形成岩石过程中,经过强白云岩化或叠加重结晶的产物。根据对巴哈马现代叠层石生物礁的研究发现,其造礁微生物群落的生命活动具有富集镁的特点,可导致碳酸盐沉积物白云石化,由此可解释为什么大量的叠层石化石都保存在白云岩中。

Grey-black thick bedded and massive stromatolite dolostones in the lower part of the Luanshigou Formation. Dolostone or dolomite rock is a carbonate sedimentary rock containing a high percentage of the mineral dolomite, $CaMg(CO_3)_2$. The most common type of dolostone is the tropical marine carbonate sediments that were dolomitized. Dolomitization means that calcium carbonate were replaced by calcium magnesium carbonate through the action of magnesium-bearing water percolating the limestone or limy mud. The study on modern stroamtolites in Bahama found that microbiota can enrich magnesium from sea water during their build up reefs and release magnesium to form high magnesian calcites easily dolomitized after their death. Perhaps it helps explain why significant amount of fossil stromatolites preserved in dolostones.

图 2.1.2 （摄影：钱迈平） Fig. 2.1.2 （Photograph by Qian Maiping）

乱石沟组下部灰黑色中薄层状细晶纹层白云岩。其灰黑色显示当时处于缺氧的沉积环境，中薄层状纹层反映当时的水体较平静，沉积物来源的多少呈较稳定的周期变化。因水体处于较停滞状态，含氧的表层水难以与下层水交流，造成下层水缺氧，加上微生物产生的有机质累积，更加剧了缺氧，使这套沉积岩偏黑色。

Grey black medium- to thin-bedded laminar fine-grained dolostones in the lower part of the Luanshigou Formation. It's grey-black coloration suggests an anoxic sedimentary environment and the medium- to thin-bedded laminar structure indicates that it was deposited periodically by suspension settling in calm water. This stagnation condition restricted exchange between the surface water dissolved oxygen and the lower water, and resulted anoxic lower waters. In addition, the rate of oxidation of accumulating organic materials by microbiota is greater than the supply of dissolved oxygen, and caused the sedimentary rock blacking.

图2.1.3 （摄影：钱迈平） Fig. 2.1.3 （Photograph by Qian Maiping）

乱石沟组下部深灰色中厚层状泥晶－细晶纹层白云岩。

Dark grey medium- to thick-bedded laminar micrite or fine-grained dolostones in the lower part of the Luanshigou Formation.

图2.1.4 （摄影：钱迈平） Fig. 2.1.4 （Photograph by Qian Maiping）

乱石沟组下部浅灰色中薄层状泥晶纹层白云岩。平缓柔曲状层理，是较平静环境沉积的水平层理在尚未固结时，因处于碳酸盐台地边缘缓坡，在重力作用下发生了轻微的滑动而变形。

Light grey medium- to thin-bedded laminar micrite dolostones in the lower part of the Luanshigou Formation. The gently folding bedding indicates that former horizontal bedding formed in calm water on a gentle slope of the carbonate platform margin slid slightly deformation by gravitational effects before consolidation.

图 2.1.5 （摄影：钱迈平）　　　　　　　　　Fig. 2.1.5　(Photograph by Qian Maiping)

乱石沟组下部灰黑色块状叠层石生物礁白云岩。这层生物礁白云岩紧密排列的，以柱状叠层石圆柱朱鲁莎叠层石（*Jurassia cylindrica*）为特征，灰黑色反映其沉积时处于缺氧的还原环境。值得注意的是，这种叠层石通常形成于光照充足的浅水氧化环境，却埋葬在还原环境沉积物里，说明这些还原环境沉积的黑色淤泥曾被一次突发的强大波浪搅起，侵入了叠层石生长区，埋葬了这些叠层石！当时究竟发生了什么？也许是大风暴，或大海啸，或大地震，或其他什么地质事件。

Grey black massive stromatolite dolostones in the lower part of the Luanshigou Formation. Note: the columnar stromatolites *Jurassia cylindrica* formed in a sunny oxygen-rich condition occurred abnormally in the grey black sediments formed in an anoxic condition. It would suggest an outburst of great waves with the black mud from an anoxic condition intruded into the stromatolite communities zone, and then covered them! What happened? Maybe it was a strong storm, or a big tsunami, or a violent earthquake, or other something.

图 2.1.6 （摄影：钱迈平）
Fig. 2.1.6 （Photograph by Qian Maiping）

乱石沟组下部深灰色中薄层状泥晶纹层白云岩，层面呈浅紫红色，泥裂构造发育。显示干燥炎热气候潮坪沉积特点，即退潮暴露出水期间，在太阳暴晒下，泥质沉积物因其表面和下面的脱水收缩不一致，而产生多边形网状开裂；其浅紫红色是暴露期间受氧化作用的结果。

Dark grey medium - to thin-bedded laminar micrite dolostones, with dessication cracks on the purplish red top layer, in the lower part of the Luanshigou Formation. It is a depositional characteristic of a tidal flat in hot and dry weather, where wet muddy sediment aerial exposure, dries up and contracts after the ebb. A strain is developed because the top layer shrinks while the material below stays the same size. When this strain becomes large enough, channel cracks form in the dried-up surface to relieve the strain. Individual cracks spread and join up, forming a polygonal, interconnected network. Its purplish red coloration suggests an oxidation on the top layer.

图 2.1.7 （摄影：钱迈平）　　　　　Fig. 2.1.7 （Photograph by Qian Maiping）

乱石沟组下部灰黑色中薄层状粉晶纹层白云岩，方解石脉发育。从方解石脉的分布及走向看，该地层在形成岩石后，至少遭受过两个方向的张性剪切力作用，形成了两组不同方向展布的雁列方解石脉。

Grey black medium - to thin-bedded laminar silty dolostones, with developed white en échelon calcite veins, in the lower part of the Luanshigou Formation. You can see en échelon veins appear as sets of short, parallel, calcite-filled lenses within the dolostone. They were caused by noncoaxial shear in at least two directions.

图 2.1.8 （摄影：钱迈平）　　　　　　　　　　Fig. 2.1.8　（Photograph by Qian Maiping）

乱石沟组下部灰黑色薄层状夹黑色硅质条带粉晶白云岩。这些硅质条带成层延展，并具有纹层，显示其当时邻近海底火山活动区。在碳酸盐沉积期间，火山时常喷出富含硅的热液夹在其中。纹层微弱起伏，是较平静环境沉积的水平层理在尚未固结前，因处于碳酸盐台地边缘斜坡，在重力作用下发生了不大的滑动而微弱变形。

Grey black thin-bedded silty dolostones interbedded with abundant black laminar chert layers in the lower part of the Luanshigou Formation. These chert layers were formed by silica-rich hydrothermal vent fluids near volcanically active places. Their gently folding bedding indicates that former horizontal bedding formed in a calm water condition on a gentle slope of the carbonate platform margin slid slightly deformation by gravitational effects before consolidation.

图2.1.9 （摄影：钱迈平） Fig. 2.1.9 （Photograph by Qian Maiping）

乱石沟组下部灰黑色中厚层状叠层石礁白云岩，波层叠层石（*Stratifera undata*）发育。

Grey black medium- to thick-bedded stromatolite dolostones with stratiform stromatolite *Stratifera undata* in the lower part of the Luanshigou Formation.

图 2.1.10 （摄影：钱迈平） Fig. 2.1.10 (Photograph by Qian Maiping)

乱石沟组下部灰黑色中厚层状纹层白云岩，黑色燧石纹层、平行层理和交错层理发育。交错层理，反映了当时潮汐作用下的碳酸盐台地浅水沉积；燧石纹层，体现了当时附近的海底火山时常在喷涌富硅热液。

Grey black medium- to thick-bedded laminar silty dolostones interbedded with black laminar chert layers, with well-developed cross-bedding, in the lower part of the Luanshigou Formation. The cross-bedding were resulted by tidal processes operating in a shallow marine carbonate platform. The laminar chert layers were formed by hydrothermal vent fluids and sediments near volcanically active places.

图 2.1.11 （摄影：钱迈平） Fig. 2.1.11 （Photograph by Qian Maiping）

乱石沟组中部灰紫色中厚层状纹层白云岩。纹层呈微弱起伏的波状，是较平静环境的沉积。其灰紫色反映当时在炎热潮湿气候下的氧化环境，碳酸盐沉积物里的菱铁矿，即碳酸铁（$FeCO_3$），很容易被氧化成褐铁矿，即含水的三氧化二铁（$Fe_2O_3 \cdot H_2O$），而呈灰紫色。

Grey purple medium- to thick-bedded laminar dolostones in the middle part of the Luanshigou Formation. The gently corrugated laminated beds formed in a calm water condition. The grey purple color indicates an oxidizing shallow-marine environment under hot and humid climate caused alterations from siderites （$FeCO_3$） to limonites （$Fe_2O_3 \cdot H_2O$） in carbonate sediments.

图 2.1.12 （摄影：钱迈平）　　　　　　　　　　Fig. 2.1.12 （Photograph by Qian Maiping）

乱石沟组中部灰紫-灰黄色块状白云质角砾岩。其角砾及胶结物都来自下伏的白云岩地层，白云岩角砾大小从毫米级到米级都有，白云质泥砂胶结，是碳酸盐台地边缘斜坡上因突发的强烈波动或震动（如风暴、火山喷发或地震）产生的重力流沉积。

Grey purple and grey yellow massive dolomitic breccias in the middle part of the Luanshigou Formation. The angular clasts in this breccia are millimeter-, centimeter- and meter-size dolostone fragments. The matrix is a mix of mud- through sand-size particles. They were both from the underlying formations and piled on the carbonate platform margin slope by gravity flow deposits when a strong disturbance or shaking (such as a storm, volcanic eruption or earthquake) occurred.

图 2.1.13 （摄影：钱迈平）　　　　　Fig. 2.1.13　（Photograph by Qian Maiping）

乱石沟组中部灰黄色中厚层状纹层白云岩。当时处于碳酸盐台地边缘斜坡，纹层因重力滑塌而揉曲变形或断裂。

Grey yellow medium- to thick-bedded laminar dolostones in the middle part of the Luanshigou Formation. The laminar dolostones with gravity gliding or gravity spreading, folding and fracturing structures formed on the carbonate platform margin slope.

图 2.1.14 （摄影：钱迈平）　　　　　　　　　　Fig. 2.1.14　(Photograph by Qian Maiping)

　　乱石沟组中部灰紫－灰黄色厚层－块状叠层石礁白云岩。叠层石柱体断断续续、东倒西歪，显示其形成于波浪作用较强的碳酸盐台地潮坪－浅滩。

Grey purple and grey yellow massive stromatolite dolostones in the middle part of the Luanshigou Formation. The intermittent and disorderly columnar stromatolites showed they formed in a tidal flat and shoal on carbonate platform under a strong wave action.

图 2.1.15 （摄影：钱迈平） Fig. 2.1.15 (Photograph by Qian Maiping)

乱石沟组中部灰紫-灰黄色厚层-块状叠层石礁白云岩。叠层石柱体直径呈周期性变化，可能反映了当时潮汐的周期性变化。

Grey purple and grey yellow massive stromatolite dolostones in the middle part of the Luanshigou Formation. The cyclic diameter variations of columnar stromatolites probably reflected the cyclic tidal variations on the carbonate platform.

图2.1.16 （摄影：钱迈平）　　　　　　Fig. 2.1.16 （Photograph by Qian Maiping）

乱石沟组中部深灰色中厚层状夹黑色燧石条带纹层白云岩。层理因重力滑塌而揉曲变形或断裂，说明其处于碳酸盐台地边缘斜坡上。硅质条带成层延展，显示当时海底火山活动频繁，不时喷出富含硅的热液夹在碳酸盐沉积中。

Dark grey medium- to thick-bedded dolostones interbedded with black laminar chert layers in the middle part of the Luanshigou Formation. The folded and fractured bedding indicates they slid deformation by gravitational effects on a slope of the carbonate platform margin. These black chert layers were submarine volcano silica-rich hydrothermal sediments.

图 2.1.17 （摄影：钱迈平）　　　　　　　　　　Fig. 2.1.17 （Photograph by Qian Maiping）

乱石沟组中部灰紫-灰黄色厚层-块状叠层石礁白云岩。叠层石由柱体向一侧大角度倾斜的圆柱朱鲁莎叠层石（*Jurassia cylindrica*）构成，显示其生长在碳酸盐台地潮坪-浅滩较强的定向水流中。

Grey purple and grey yellow thick-bedded and massive stromatolite dolostones in the middle part of the Luanshigou Formation. The stromatolite bioherms composed mainly of *Jurassia cylindrica* columns which incline to one side in a large angle and suggested they formed in a strong directional flow on a tidal flat or shoal of the carbonate platform.

图 2.1.18 （摄影：钱迈平）　　　　　　　　　　Fig. 2.1.18　（Photograph by Qian Maiping）

乱石沟组中部紫红色厚层－中厚层状纹层白云岩。紫红色纹层呈水平层理，是在炎热湿润气候下，在较平静的浅水泥坪氧化环境里，由水中细小的泥质悬浮物沉积形成。

Purple red thick and medium- to thick-bedded laminar dolostones in the middle part of the Luanshigou Formation. The lamination developed in fine grained sediment when fine grained particles settled, which happened in a quiet shallow water mudflat, where the tide creates cyclic differences in sediment supply. The purple red color indicates an oxidizing environment under hot and humid climate.

图 2.1.19 （摄影：钱迈平）　　　　　　　　　　Fig. 2.1.19 （Photograph by Qian Maiping）

乱石沟组中部灰紫-灰黄色厚层-块状纹层白云岩夹层状叠层石。纹层呈水平层理，层状叠层石类型是波层叠层石（*Stratifera undata*），其基本层不规则地向上凸起，形成一系列隆丘。显示在当时较平静的水下氧化环境里，微生物席因光合作用需要，向上趋光性凸起。

Grey purple and grey yellow thick-bedded and massive laminar dolostones interbedded with stratiform stromatolites in the middle part of the Luanshigou Formation. Their colors and structures mean they were deposited in a relatively quiet shallow-water oxidizing environment, where the stratiform stromatolite *Stratifera undata* biostromes with many photokinetic mounds for photosynthesis.

图 2.1.20 （摄影：钱迈平）　　　　　　　　　　Fig. 2.1.20 （Photograph by Qian Maiping）

乱石沟组中部灰紫－灰黄色中厚层状叠层石礁白云岩。叠层石以喀什喀什叠层石（*Kussiella kussiensis*）为主，其柱体紧密排列，当时处于碳酸盐台地潮坪－浅滩环境。

Grey purple and grey yellow medium- to thick-bedded stromatolite dolostones in the middle part of the Luanshigou Formation. The stromatolite bioherms were built up mainly by *Kussiella kussiensis* columns formed on a tidal flat of the carbonate platform.

图 2.1.21 (摄影:钱迈平)　　　　　　　　　Fig. 2.1.21　(Photograph by Qian Maiping)

这些喀什喀什叠层石(*Kussiella kussiensis*),因生长在定向水流中,叠层石柱体向一侧倾斜。

These *Kussiella kussiensis* columns incline to one side and show they formed in a directional flow.

图 2.1.22 （摄影：钱迈平） Fig. 2.1.22 (Photograph by Qian Maiping)

乱石沟组中部灰紫—灰黄色厚层—块状叠层石礁白云岩。叠层石以柱状的圆柱朱鲁莎叠层石（*Jurassia cylindrica*）为主，叠层石柱体紧密排列。

Grey purple and grey yellow thick-bedded and massive stromatolite dolostones in the middle part of the Luanshigou Formation. The stromatolite bioherms composed mainly of *Jurassia cylindrica* columns arranged in tightly parallel.

图 2.1.23 （摄影：钱迈平） Fig. 2.1.23 （Photograph by Qian Maiping）

乱石沟组中部灰紫－灰黄色厚层－块状叠层石礁白云岩。叠层石以分叉柱状的育卡贝加尔叠层石（*Baicalia unca*）为主，当时处于碳酸盐台地潮坪－浅滩。

Grey purple and grey yellow thick-bedded and massive stromatolite dolostones in the middle part of the Luanshigou Formation. The stromatolite bioherms were built up mainly by *Baicalia unca* columns formed on a tidal flat-shoal of the carbonate platform.

图 2.1.24
(摄影:钱迈平)

Fig. 2.1.24
(Photograph by Qian Maiping)

乱石沟组中部灰紫色中厚层状纹层白云岩。褶皱扭曲的纹层说明,平静水下沉积形成的纹理在尚未固结前,因处于碳酸盐台地斜坡,在重力作用下发生滑动而变形。

Grey purple medium- to thick-bedded laminar dolostones in the middle part of the Luanshigou Formation. The lamination formed in a quiet shallow water on mudflat slop occurred deformation resulted from gravitational effects before they consolidated.

图 2.1.25
(摄影：钱迈平)

Fig. 2.1.25
(Photograph by Qian Maiping)

  乱石沟组上部，先后出现了3层浅紫色中薄－中厚层状燧石角砾岩（俗称"宝石砾岩"），夹在紫红－浅紫色中薄－中厚层状纹层白云岩中。色彩斑斓的燧石小角砾来自于下面白云岩地层的热液沉积燧石夹层，紫红、灰白、灰黑及黑色的都有，砾石大小呈毫米－厘米级混杂，分布不均匀，大多呈次棱角状，其胶结物为白云岩。由此判断：是多次突发的强烈海浪将已沉积的夹燧石薄层的白云岩打碎，形成高密度的水下碎屑流，汹涌而下，快速堆积。这很可能与大风暴带来的强浪、大地震或海底火山喷发引起的海啸有关。

  Three layers of light purple medium- to thin-bedded and medium- to thick-bedded chert breccias (commonly known as "gem breccias") in the purple-red and light purple medium- to thin-bedded and medium- to thick-bedded lamellar dolostones in the upper part of the Luanshigou Formation. Colourful, including purple-red, grey-white, grey-black and black etc., gravels in the breccias are from underlying hydrothermal chert interlayers in the dolostones. These gravels are millimeter- to centimeter-sized and they are mostly angular to subangular fragments of rocks. They distributed unevenly in a finer grained dolomitic groundmass which were produced by mass wasting. The breccias were formed by high-density submarine debris flows resulted likely from the strong waves by large stroms or the tsunamis by earthquakes or volcanoes.

图 2.1.26 （摄影：钱迈平）　　　　　　　　　　Fig. 2.1.26　(Photograph by Qian Maiping)

　　乱石沟组上部紫红－浅紫色中薄－中厚层状纹层白云岩。岩石的色泽和水平纹层，显示其沉积在浪基面以下的较平静浅水环境。

　　The purple red and light purple medium- to thin-bedded and medium- to thick-bedded lamellar dolostones in the upper part of the Luanshigou Formation. The purple color and horizontally laminated beds indicate a calm shallow-water oxidizing environment below wave base.

图 2.1.27 （摄影：钱迈平） Fig. 2.1.27 （Photograph by Qian Maiping）

乱石沟组上部紫红－浅紫色中厚层状叠层石礁白云岩。叠层石以波层叠层石（*Stratifera undata*）为主，显示当时处于适合光合作用的光照带以上较平静浅水环境。

The purple red and light purple medium- to thick-bedded stromatolite dolostones in the upper part of the Luanshigou Formation. The stromatolite biostromes were built up mainly by *Stratifera undata* in a relatively quiet shallow-water environment which was suitable for photosynthesis.

图 2.1.28　（摄影：钱迈平）　　　　　　　　　Fig. 2.1.28　（Photograph by Qian Maiping）

乱石沟组上部紫红中薄层状纹层白云岩。层面可见链形波痕构造，显示当时处于热带潮坪颤动波浪弱水流环境。

Catenary ripple marks in purple red medium- to thin-bedded laminar dolostones in the upper part of the Luanshigou Formation. They indicated a tropic tidal flat environment with weak currents where water motion was dominated by wave oscillations.

图 2.1.29 （摄影：钱迈平）　　　　　　　　Fig. 2.1.29　(Photograph by Qian Maiping)

乱石沟组上部浅紫色厚层一块状含角砾白云岩。下部含砾石较多，往上砾石变少，最上部纹层发育。显示了碳酸盐台地边缘斜坡夹带砾石等沉积物的重力流从汹涌而下到逐渐平息的沉积过程。

Light purple thick-bedded to mass breccia-bearing dolostones in the upper part of the Luanshigou Formation. There was the graded bedding that gravels became less and less passing upwards into laminar marls in the top. It was the result of flood waters carrying a pulse of sediment into an underwater landslide on a slope of the carbonate platform.

图 2.1.30 （摄影：钱迈平）　　　　　　　　　　Fig. 2.1.30　（Photograph by Qian Maiping）

砾石来自下面地层的热液沉积燧石夹层，浅紫、灰白及灰黑色的都有，分选和磨圆度差，分布不均匀，反映了水下碎屑流快速堆积的特征。

These light purple, grey white and grey black gravels in the breccias were from the chert interlayers of underlying strata. These unsorted, poorly rounded and unevenly distributed gravels were deposited from submarine debris flows resulted likely from the strong tidal waves.

图 2.1.31 （摄影：钱迈平）　　　　　　　　Fig. 2.1.31 （Photograph by Qian Maiping）

随着海啸余波的逐渐减弱，水中悬浮物呈周期性沉淀，形成沉积纹层。一些悬浮的小砾石因比白云质泥砂重，先沉淀下来，夹在下部纹层中；而后沉淀的上部纹层就没有砾石，都是白云质泥砂。

As the aftereffect of tidal waves weakened, pulse waned, water lost velocity gradually, and suspended particles were deposited. Coarsest materials containing many gravels settled first, medium next, then fine without any gravels.

图 2.1.32 （摄影：钱迈平）　　　　　　　　　　Fig. 2.1.32 　（Photograph by Qian Maiping）

乱石沟组上部紫红色中厚层状叠层石礁白云岩，叠层石以波层叠层石（*Stratifera undata*）为主，显示当时处于适合光合作用的光照带以上较平静浅水环境。

Purple red medium- to thick-bedded stromatolite dolostones in the upper part of the Luanshigou Formation. The stromatolite biostromes were built up mainly by *Stratifera undata* in a relatively quiet shallow-water environment which was suitable for photosynthesis.

图 2.1.33 （摄影：钱迈平） Fig. 2.1.33 （Photograph by Qian Maiping）

乱石沟组上部浅紫色中厚层状泥晶纹层白云岩。是热带气候碳酸盐台地潮下带沉积。

Light purple medium- to thick-bedded lamellar micritic dolostones in the upper part of the Luanshigou Formation. It indicated a tropic subtidal environment of a carbonate platform.

图 2.1.34　（摄影：钱迈平）　　　　　　　　Fig. 2.1.34　（Photograph by Qian Maiping）

乱石沟组上部浅紫色厚层－块状叠层石礁白云岩。叠层石以波层叠层石(*Stratifera undata*)为主,显示当时处于较平静浅水环境。

Light purple thick-bedded to massive stromatolite dolostones in the upper part of the Luanshigou Formation. The stromatolite biostromes built up mainly by stratiform stromatolites *Stratifera undata* in a relatively calm shallow-water environment.

图 2.1.35
(摄影:钱迈平)

Fig. 2.1.35
(Photograph by Qian Maiping)

神农顶国家大气背景监测站下,出露乱石沟组上部地层,自下而上可见:乱石沟组上部,紫红色薄层泥质状纹层白云岩与浅紫色薄－中薄层状白云质泥岩不等厚互层,水平层理与重力滑揉层理不等厚交替出现,重力滑揉、断裂及注射脉构造发育。是干旱－湿润季节交替的炎热气候下,萨布哈盐坪斜坡沉积,因不时地受扰动(风暴及潮汐)或振动(火山喷发及地震),造成沉积层不时地发生揉曲、断裂并向下坡方向滑动。

The outcrop of the upper part of the Luanshigou Formation at near the National Atmospheric Background Station in Shennongding in an ascending order: in the upper part of the Luanshigou Formation, alternating beds of the medium - to thin-bedded purple red muddy laminar dolostone and light purple dolomitic marlstone, with gravity gliding or gravity spreading, folding, fracturing and injection dyke structures. It suggested that the alternating carbonate and marl sediments on a slope in the sabkha environment alternated between dry and wet seasons were driven to deformation by gravity and fractured when a disturbance or shaking (such as a storm, volcanic eruption or earthquake) occurred.

图 2.1.36 （摄影：钱迈平）　　　　　　　　　Fig. 2.1.36 （Photograph by Qian Maiping）

　　乱石沟组上部紫红色泥质纹层白云岩与浅紫色白云质泥岩不等厚互层。可见重力作用下的滑动、揉曲、断裂及注射脉构造。

Alternating beds of purple red muddy laminar dolostone and light purple dolomitic marlstone in the upper part of the Luanshigou Formation. Noted the gravity gliding, folding, fracturing and injection dyke structures.

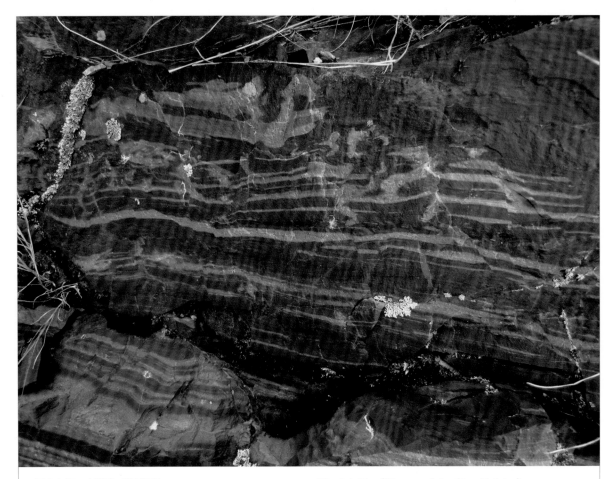

图2.1.37 （摄影：钱迈平）　　　　　　　　　　Fig. 2.1.37　（Photograph by Qian Maiping）

可见下层浅紫色白云质泥岩注射脉沿裂隙分布在上层紫红色泥质白云岩内。这是因为当沉积物尚未固结，而上面的沉积层密度大于下面的沉积层密度时，一旦发生扰动（风暴及潮汐）或振动（火山喷发及地震），下面密度较小而饱含水分的沉积层就会沿扰动或振动产生的裂隙进入上面密度较大而不透水的沉积层。

The injection dykes from underlying light purple dolomitic marlstones rose into the purple-red muddy dolostones. It shows the dense, impermeable carbonate beds were fractured, when a disturbance or shaking (such as a storm, volcanic eruption or earthquake) occurred, meanwhile the underlying unconsolidated, loosely packed and water-saturated marls would rise through the fissures and flow into the carbonate beds, as a result of difference in density. This is a soft-sediment deformation in unlithified sediments.

图 2.1.38 （摄影：钱迈平） Fig. 2.1.38 （Photograph by Qian Maiping）

乱石沟组上部浅紫色中薄层状白云质泥岩中的揉曲变形构造。也是在沉积物尚未固结前，因扰动或振动导致的软沉积物变形构造。

In the upper part of the Luanshigou Formation, folding structures in the light purple dolomitic marlstone were soft-sediment deformations in unlithified and water-saturated sediments. It was resulted from a disturbance or shaking (a storm, volcanic eruption or earthquake).

图 2.1.39 （摄影：钱迈平） Fig. 2.1.39 (Photograph by Qian Maiping)

乱石沟组上部紫红色泥质纹层白云岩与浅紫色白云质泥岩不等厚互层。可见重力滑动、揉曲构造。

Alternating beds of the medium - to thin-bedded purple red muddy laminated dolostone and light purple dolomitic marlstone in the upper part of the Luanshigou Formation. Note the gravity gliding and folding structures.

图2.1.40 （摄影：钱迈平）　　　　　　　　Fig. 2.1.40 （Photograph by Qian Maiping）

乱石沟组上部紫红色泥质纹层白云岩与浅紫色白云质泥岩不等厚互层。重力滑动方向指向斜坡下的海盆。

Alternating beds of the medium- to thin-bedded purple-red muddy laminated dolostone and light purple dolomitic marlstone with gravity gliding and folding structures in the upper part of the Luanshigou Formation.

图2.1.41 （摄影：钱迈平） Fig. 2.1.41 (Photograph by Qian Maiping)

乱石沟组上部浅紫色中薄层状白云质叠层石泥灰岩。叠层石以波层叠层石（*Stratifera undata*）为主，形成在萨布哈盐坪的干热季节。

Light purple the medium- to thin-bedded dolomitic stromatolite marlstone comprised mainly *Stratifera undata* in the upper part of the Luanshigou Formation. It suggested that they formed in a hot and dry season in a sabkha environment.

图 2.1.42 （摄影：钱迈平） Fig. 2.1.42 （Photograph by Qian Maiping）

乱石沟组上部紫红中薄层状纹层白云岩，层面可见蜿蜒的波痕，显示当时处于萨布哈盐坪浅水，潮汐水流的波浪作用在盐坪表面留下的痕迹。

Wave sinuous ripple marks in the purple red muddy laminated dolostone of the upper part of the Luanshigou Formation. It created by unidirectional flow in water shallower than wave base.

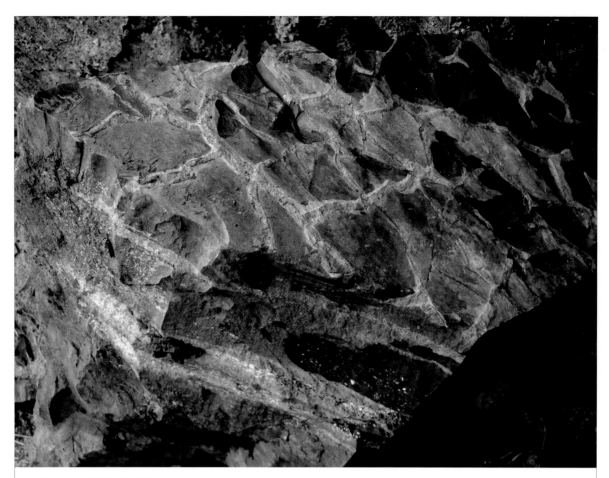

图 2.1.43 （摄影：钱迈平）　　　　　　　　　　Fig. 2.1.43　（Photograph by Qian Maiping）

乱石沟组上部紫红薄层状粉晶白云岩，层面可见泥裂构造。是当时其处于萨布哈盐坪，在暴露出水期间，被阳光暴晒，沉积层表面干裂留下的痕迹。

Mud cracks in the purple red thin-bedded fine-crystalline dolostone of the upper part of the Luanshigou Formation. It shows a hot and dry sabkha environment.

乱石沟组地层沉积于碳酸盐台地浅水斜坡，自下而上的岩石色彩明显分为三部分：下部深灰－灰黑色，中部灰紫色，上部紫红－浅紫色。其形成过程如下：

起初，这里因邻近海底地壳活动带，而且附近有障蔽海湾或较深海盆存在。台地浅水区的水流通畅，是光照良好的氧化环境，进行光合作用的蓝细菌形成许多叠层石生物礁；障蔽海湾或较深海盆的底层水体较停滞，形成缺氧的还原环境，富含硫化氢，沉积物呈黑色。由于时常的海底地壳活动引发火山喷发及地震，将附近障蔽海湾或较深海盆的底层黑色沉积物大量搅起，形成高密度混浊泥浆水团，侵入潮坪浅水区，导致部分叠层石生物礁生长区的碳酸盐沉积形成的岩石呈现灰黑色。此外，频繁的海底火山活动提供的富硅质热液，夹在碳酸盐中沉积，在一些白云岩中形成大量的燧石夹层。

The Luanshigou Formation formed in shallow water on a slope of the carbonate platform. The rocks clearly show three different colors from bottom to top: dark grey and black in the lower part, grey purple in the middle part, and purple red and light purple in the upper part.

Dark grey and black stromatolite dolostones and chert banded dolostones indicted the platform initially was perhaps located near a seafloor seismically and volcanically active area, meanwhile near a barrier bay or sea basin where the relatively deep and stagnant water riched in highly toxic hydrogen sulfide and deposited black sediments. It would suggest underwater earthquakes and volcano eruptions had frequently occurred and it could cause outbursts of great waves with black sediments from an anoxic condition in a barrier bay or sea basin to intrude frequently into the stromatolite communities zone and deposit frequently silica-rich sediments from hydrothermal vent fluids

随着海水逐渐变浅,台地光照充足,更加有利于微生物席进行光合作用,叠层石生物礁也越来越发育,构成规模宏大的大堡礁,氧气产量明显增加,沉积形成的岩石因其所含氧化的三价铁而趋显红色。起初呈现灰紫色,当海水进一步变浅,并处于旱季－雨季交替的热带沙漠气候下,沉积形成的岩石呈现紫红、浅紫色互层,波痕、泥裂和石盐假晶等构造开始出现,显示萨布哈盐坪特征。

由于这里一直处于台地边缘斜坡,重力流、重力滑动及重力扩张频繁。

produced by submarine volcanisms.

As the sea water became shallow gradually, the carbonate platform became sunny and suited well for development of photosynthetic microbial mats. Stromatolite bioherms and biostromes became bloom and built great barrier reefs, and produced more oxygen caused to that oxidizing conditions prevailed in the sediments during sedimentation of the rocks, and it was reddened by impregnation of hematite. At first it was grey purple, subsequently alternating purple-red and light purple as sea water became shallower and shallower under a tropical desert climate with alternating rainy and dry seasons, and became eventually a sabkha tidal flat with characteristic sedimentary structures such as ripple marks, mud cracks and halite pseudomorphs etc.

The gravity flows, gravity slids and gravity spreading occurred frequently due to steepness of the slope of the carbonate platform.

2.2 大窝坑组

大窝坑组厚224 m,以浅灰色泥晶－粉晶白云岩为主,发育多层叠层石生物礁白云岩。

沉积构造以水平层理为主,具条带状构造、波状层理、斜层理。

与下伏地层乱石沟组顶部浅紫色含砾白云岩整合接触。

2.2 DAWOKENG FORMATION

Dawokeng Formation is 224 m thick and is mainly consisted of light grey muddy and micritic dolostones including several stromatolite biostrome-bioherm dolo-stones.

The sedimentary structures in this formation are dominated by banded horizontal bedding, along with cross bedding and ripple marks.

This formation is in conformable contact with the underlying light purple breccia-bearing dolostone in the top of the Luanshigou Formaion.

图 2.2.1 （摄影：钱迈平）　　　　　　　　　Fig. 2.2.1　（Photograph by Qian Maiping）

神农顶国家大气背景监测站附近，出露大窝坑组底部，为浅灰色中薄层状叠层石礁白云岩。叠层石类型是波层叠层石（*Stratifera undata*），形成于碳酸盐台地潮坪上部。

Light grey medium- to thin-bedded stromatolite dolostones in the bottom of the Dawokeng Formation near Shennongjia National Air Background Monitoring Station at Shennongding. The stratiform stromatolite *Stratifera undata* biostromes formed on the upper tidal flat on the carbonate platform.

图 2.2.2　（摄影：钱迈平）　　　　　　　　Fig. 2.2.2　（Photograph by Qian Maiping）

神农顶－凉风垭公路沿线出露的地层剖面：

大窝坑组底部浅灰色中薄层状叠层石礁白云岩。叠层石类型是波层叠层石（*Stratifera undata*），形成于碳酸盐台地潮坪上部。

Along Shennongding-Liangfengya highway:

Light grey medium- to thin-bedded stromatolite dolostones in the bottom of the Dawokeng Formation. The stratiform stromatolite *Stratifera undata* biostromes formed on the upper tidal flat on the carbonate platform.

图 2.2.3 （摄影：钱迈平）　　　　　　　　　Fig. 2.2.3 （Photograph by Qian Maiping）

大窝坑组下部浅灰色中薄层状纹层白云岩。纹层微弱起伏，是较平静环境沉积的水平层理。

Light grey medium- to thin-bedded laminar dolostones in the lower part of the Dawokeng Formation. Their gently undulating bedding indicates a relatively calm underwater environment.

图 2.2.4 （摄影：钱迈平）　　　　　　　　　　　Fig. 2.2.4　（Photograph by Qian Maiping）

大窝坑组下部浅灰色厚层状叠层石礁白云岩。叠层石类型是波层叠层石（*Stratifera undata*），形成于碳酸盐台地潮坪上部较平静环境。其微生物席因光合作用，生长出一个个趋光性丘状隆起。这些隆起呈倾斜状，而在平静水下叠层石通常垂直海平面向上生长，由此测量其隆起方向与层面夹角，推算当时海底坡度在 5°—10°之间。

Light grey thick-bedded stromatolite dolostones in the lower part of the Dawokeng Formation. The stratiform stromatolite *Stratifera undata* biostromes formed in a relatively calm underwater environment on the carbonate platform. Tilting photokinetic mounds for photosynthesis of the stromatolites indicates the slopes angles on the platform were between 5° and 10°.

图2.2.5 （摄影：钱迈平）　　　　　　　　Fig. 2.2.5 （Photograph by Qian Maiping）

大窝坑组下部浅灰色厚层状叠层石礁白云岩。叠层石礁由波层叠层石（*Stratifera undata*）及简单包心菜叠层石（*Cryptozoon haplum*）构成，形成于碳酸盐台地潮坪中－上部。

Light grey thick-bedded stromatolite dolostones in the lower part of the Dawokeng Formation. The stratiform stromatolite *Stratifera undata* biostromes and domed stromatolite *Cryptozoon haplum* bioherms formed on the middle to upper tidal flat on the carbonate platform.

图 2.2.6 （摄影：钱迈平） Fig. 2.2.6 （Photograph by Qian Maiping）

大窝坑组下部浅灰色厚层状叠层石礁白云岩。叠层石礁由波层叠层石（*Stratifera undata*）构成，形成于碳酸盐台地潮坪上部。

Light grey thick-bedded stromatolite dolostones in the lower part of the Dawokeng Formation. The stratiform stromatolite *Stratifera undata* biostromes formed on the upper tidal flat on the carbonate platform.

图 2.2.7 （摄影：钱迈平） Fig. 2.2.7 （Photograph by Qian Maiping）

大窝坑组下部灰黑色中厚层状纹层白云岩。纹层微弱起伏，是较平静环境沉积的层理。

Grey black medium- to thick-bedded laminar dolostones in the lower part of the Dawokeng Formation. Their gently undulating bedding indicates a relatively calm underwater environment.

图2.2.8 （摄影：钱迈平） Fig. 2.2.8 （Photograph by Qian Maiping）

大窝坑组下部灰黑色厚层－块状叠层石礁白云岩。叠层石礁由树桩圆柱叠层石（*Colonnella cormosa*）构成，形成于碳酸盐台地潮下带水稍深处，并受火山作用影响有所硅化。

Grey black thick-bedded to massive stromatolite dolostones in the lower part of the Dawokeng Formation. The column stromatolite *Colonnella cormosa* formed in the subtidal zone on the carbonate platform. These stromatolites were in a relatively deep underwater environment and were slightly silicified by volcanisms.

图 2.2.9 （摄影：钱迈平）　　　　　　　　　Fig. 2.2.9　（Photograph by Qian Maiping）

大窝坑组下部浅灰色厚层状叠层石礁白云岩。叠层石礁由波层叠层石（*Stratifera undata*）及大大小小的简单包心菜叠层石（*Cryptozoon haplum*）构成，形成于碳酸盐台地潮下带上部及潮间带。

Light grey thick-bedded stromatolite dolostones in the lower part of the Dawokeng Formation. The stratiform stromatolite *Stratifera undata* biostromes and domed stromatolite *Cryptozoon haplum* bioherms formed on the shallow subtidal and lower intertiadal zones on the carbonate platform.

图 2.2.10　（摄影：钱迈平）　　　　　　　　　　　Fig. 2.2.10　（Photograph by Qian Maiping）

大窝坑组下部深灰色厚层状叠层石礁白云岩。叠层石礁由波层叠层石（*Stratifera undata*）构成，形成于碳酸盐台地潮坪中—上部。

Dark grey thick-bedded stromatolite dolostones in the lower part of the Dawokeng Formation. The stratiform stromatolite *Stratifera undata* biostromes formed on the mid- to upper tidal flat on the carbonate platform.

图 2.2.11 （摄影：钱迈平）　　　　　　　　　　Fig. 2.2.11 （Photograph by Qian Maiping）

大窝坑组下部深灰色厚层状叠层石礁白云岩。叠层石礁由波层叠层石（*Stratifera undata*）构成，形成于碳酸盐台地潮坪上部。

Dark grey thick-bedded stromatolite dolostones in the lower part of the Dawokeng Formation. The stratiform stromatolite *Stratifera undata* biostromes formed on the upper tidal flat on the carbonate platform.

图 2.2.12 （摄影：钱迈平） Fig. 2.2.12 （Photograph by Qian Maiping）

大窝坑组下部深灰色厚层状叠层石礁白云岩。叠层石礁由波层叠层石（*Stratifera undata*）构成，形成于碳酸盐台地潮坪上部。

Dark grey thick-bedded stromatolite dolostones in the lower part of the Dawokeng Formation. The stratiform stromatolite *Stratifera undata* biostromes formed on the upper tidal flat on the carbonate platform.

图 2.2.13 （摄影：钱迈平） Fig. 2.2.13 （Photograph by Qian Maiping）

大窝坑组下部灰紫色块状白云质角砾岩。是碳酸盐台地遭受强烈扰动（如风暴、火山喷发或地震）引发的重力流在边缘斜坡的堆积。

Grey purple massive dolomitic breccias in the lower part of the Dawokeng Formation. It was piled on the carbonate platform margin slope by gravity flow deposits when a strong disturbance or shaking (such as a storm, volcanic eruption or earthquake) occurred.

图 2.2.14　（摄影：钱迈平）　　　　　　　　Fig. 2.2.14　（Photograph by Qian Maiping）

　　大窝坑组上部浅灰色厚层-块状叠层石礁白云岩。叠层石以锥叠层石（*Conophyton*）为主，形成于碳酸盐台地潮下带。

Light grey thick-bedded massive to stromatolite dolostones in the upper part of the Dawokeng Formation. The bioherms consisted primarily of conical stromatolites *Conophyton* formed in a subtidal zone on the carbonate platform.

图 2.2.15 （摄影：钱迈平） Fig. 2.2.15 (Photograph by Qian Maiping)

这些锥叠层石（*Conophyton*）形成于潮下带较深水处，那里光照比浅水处弱，使进行光合作用的微生物席极力向上趋光生长，并大多集中在最顶端，造成微生物席急剧突起形成一个个尖锥状，所以叠层石柱体高而直，基本层在顶尖处加厚。

These stromatolites *Conophyton* had a conical growth habit in a deeper subtidal environment where sunlight was so weak that photosynthetic microbial mats had done their utmost to grow upward due to phototactic behavior, hence these high and straight conical stromatolites with thickened top of dark laminae were formed.

图 2.2.16 （摄影：钱迈平）　　　　　　　　　　Fig. 2.2.16　（Photograph by Qian Maiping）

　　大窝坑组顶部浅灰色中薄层状与浅黄色中厚层状纹层白云岩互层。纹层微弱平缓起伏，是潮下带较平静环境沉积的水平层理。

Cyclically alternating light grey medium- to thin-bedded and light yellow medium- to thick-bedded laminar dolostones in the upper part of the Dawokeng Formation. Their gently undulating laminae indicates a relatively calm subtidal environment.

图 2.2.17 （摄影：钱迈平）　　　　　　　　　Fig. 2.2.17　（Photograph by Qian Maiping）

在大窝坑组最顶部至少夹有7层条带状玉髓层，显示当时这里的古地理位置邻近海底火山活动区域，火山排放的硅质热液沉积物夹在碳酸盐沉积物中。

The laminar dolostones interbedded with at least seven layers of chrysoprase in the top of the Dawokeng Formation. These chrysoprase layers formed by hydrothermal fluids near an active submarine volcano.

图 2.2.18 （摄影：钱迈平） Fig. 2.2.18 (Photograph by Qian Maiping)

这些条带状玉燧层因含有金属镍而呈浅绿—碧绿色，非晶质结构，单层厚度在 40－140 mm 不等，沉积纹层发育，纹层微弱平缓起伏，显示较平静的低能水动力环境特征。

These chrysoprase layers are light green and aquamarine chalcedony that derives its color from nickel, ranging from nearly opaque to nearly transparent. The thickness ranges from 40 mm to 140 mm, and their gently undulating laminae indicates a relatively calm environment.

大窝坑组地层沉积于碳酸盐台地边缘浅水斜坡,叠层石生物礁白云岩发育,时有重力流形成的砾岩层。根据叠层石类型特征,可将该组分为下部和上部两部分。下部发育大量的层叠层石(*Stratifera*)和包心菜叠层石(*Cryptozoon*),显示其处于海水较浅的潮坪中、上部;上部发育锥叠层石(*Conophyton*),说明海水有所加深,已处于潮坪下部。顶部出现多层条带状玉燧层,反映其古地理位置已临近海底火山活动区域。

The Dawokeng Formation was sedimented in a shallow water on the carbonate platform margin slope, where there were developed stromatolite biostromes and bioherms, as well as some gravity flow deposits. Plenty of stratiform stromatolites *Stratifera* and domed stromatolites *Cryptozoon* in the lower part of the Dawokeng Formation indicates a middle to upper tidal flat. Well developed conical stromatolites *Conophyton* in the upper part of this formation showed the water had deepened and the habitat had been transformed into a subtidal flat. Several chrysoprase layers in the top of this formation suggested it had been located near an active submarine volcano.

2.3 矿石山组

矿石山组厚442 m,以各种白云岩为主,发育多层叠层石生物礁白云岩。该组下部为深灰-灰黑色铁质白云岩及赤铁矿层,因此被命名为矿石山组。

沉积构造以水平层理为主,具条带状构造。此外,自碎角砾、网状注射脉、液化侵入及变形等地震沉积构造发育。

与下伏地层大窝坑组顶部浅灰色中薄层状纹层白云岩整合接触。

2.3 KUANGSHISHAN FORMATION

Kuangshishan Formation is 442 m thick and are mainly consisted of various dolostones including several stromatolite bioherm-biostrome dolostone beds.

The sedimentary structures in this formation are dominated by banded horizontal bedding, along with earthquake-induced autoclastic breccias, anastomosing-injection dykes, liquefied disorganizations and intrusions etc.

This formation is in conformable contact with the underlying light gray middle-thin bedded laminar dolostones in the top of the Dawokeng Formation.

图 2.3.1
（摄影：钱迈平）

Fig. 2.3.1
(Photograph by Qian Maiping)

矿石山组底部灰黑色薄—中薄层状页岩。

Grey black thin-bedded and medium-to thin-bedded shales in the bottom of the Kuangshishan Formation.

图 2.3.2
（摄影：钱迈平）

Fig. 2.3.2
(Photograph by Qian Maiping)

细腻的泥质结构、水平层理沉积构造，反映其形成于浪基面以下较平静的低能水动力环境。

These shales composed of fine mud-sized particles and were deposited in horizontal or near-horizontal layers formed in a low energy hydrodynamic and relatively quiet environment below wave base.

图2.3.3 （摄影：钱迈平） Fig. 2.3.3 （Photograph by Qian Maiping）

灰黑色显示其沉积于水体较停滞的缺氧海底，说明此处当时海水已有所加深，并出现分层，即表层富氧水与下层缺氧水相互不交流。

The grey black coloration suggests that the shales deposit in a relatively deep and anoxic water condition uncommunicated with oxic surface water.

图 2.3.4 （摄影：钱迈平） Fig. 2.3.4 （Photograph by Qian Maiping）

矿石山组下部灰黑色厚层－块状铁质纹层白云岩。

Grey black thick- to massive bedded lamellar ferruginous dolostones in the lower part of the Kuangshishan Formation.

图2.3.5 （摄影：钱迈平） Fig. 2.3.5 (Photograph by Qian Maiping)

纹层颜色深浅交替，含铁量也相应交替变化，反映了此处化学沉积环境的周期变化。

The alternating dark and light colored layers indicates alternating rich and poor iron contents of deposits resulted from an alternating change chemical environment.

图 2.3.6 （摄影：钱迈平）　　　　　　　　　Fig. 2.3.6 （Photograph by Qian Maiping）

含铁纹层起伏平缓，显示其沉积于较平静的低能水动力环境，处于浪基面之下较深的海底。

The gently corrugated laminated ferruginous beds formed in a low energy hydrodynamic and relatively quiet environment below the wave base.

图 2.3.7 （摄影：钱迈平） Fig. 2.3.7 (Photograph by Qian Maiping)

矿石山组下部深灰色块状叠层石生物礁白云岩。叠层石生物礁由大型的加尔加诺锥叠层石（Conophyton garganicum）构成，这种叠层石通常形成于碳酸盐台地潮下带较深海底，但其深度仍在光照带之内，有足够的阳光供微生物进行光合作用。现代海洋学研究发现，即使是很清澈的水，大部分可见光谱在进入水下不到 10 m 就已被吸收掉了，极少能穿透到水下 150 m。另外，这些大型叠层石的柱体之间充满风浪打碎的岩石碎屑，至少说明其生长深度不会超过浪基面。而现代海洋学研究发现，通常好天气下浪基面位于水下 5－15 m，而风暴期间浪基面位于水下 15－40 m。由此推测，尽管这些大型锥叠层石形成在较深水下，但深度不会超过 40 m。

Dark gray massive stromatolite dolostones in the lower part of the Kuangshishan Formation. The bioherm consisted of big conical stromatolites *Conophyton garganicum* formed in a subtidal deep euphotic zone on the carbonate platform where there was still enough sunlight for photosynthesis. Most of the visible light spectrum is absorbed within 10 m of the water's surface, and almost none penetrates below 150 m of water depth, even when the water is very clear. The abundance of rock fragments or gravels, on the other hand, between columns of these big conical stromatolites show they formed in a turbulent condition above the wave base. The wave base, in physical oceanography, is the maximum depth at which a water wave's passage causes significant water motion. The wave base typically is a sea depth of 5 to 15 m below sea level during a fair weather or 15 to 40 m below sea level during a storm. Therefore, it is speculated that these stromatolites formed in a relative deep-water setting above 40 m of water depth.

图 2.3.8 （摄影：钱迈平）　　　　　　　Fig. 2.3.8　（Photograph by Qian Maiping）

这些叠层石个体呈不分枝的圆锥柱状,大多高度超过 1 m,直径 0.4—0.6 m,相互紧密平行排列,间距不超过 0.05 m。

These big non-branching columnar stromatolites are more than 1 m high, 0.4—0.6 m in diameter, in vertically parallel arrangement, no more than 0.05 m apart.

图2.3.9 （摄影：钱迈平）　　　　　　　　　Fig. 2.3.9 （Photograph by Qian Maiping）

　　矿石山组中部浅灰色块状叠层石生物礁白云岩。叠层石生物礁由波层叠层石（*Stratifera undata*）构成，这种叠层石通常形成于碳酸盐台地潮坪。

Light grey massive stromatolite dolostones in the middle part of the Kuangshishan Formation. The bioherms consisted of stratiform stromatolite *Stratifera undata* formed on a tidal flat on the carbonate platform.

图2.3.10　（摄影：钱迈平）　　　　　　　　　　Fig. 2.3.10　（Photograph by Qian Maiping）

其叠层石基本层混乱变形，上拱破裂，是因地震而导致沉积物液化并高压上涌造成的结果。

The laminae of these stromatolites are disorganized, ripped and intruded by the earthquake-induced overpressure liquefied sediments.

图 2.3.11 （摄影：钱迈平）　　　　Fig. 2.3.11 （Photograph by Qian Maiping）

矿石山组中部浅灰色块状叠层石生物礁白云岩。叠层石生物礁由瘤通古斯叠层石（*Tungussia nodosa*）构成，形成于碳酸盐台地高能水动力潮间带。

Light grey massive stromatolite dolostones in the middle part of the Kuangshishan Formation. The bioherms consisted of frequent divergent branching columnar stromatolite *Tungussia nodosa* formed in a high hydrodynamic energy intertidal zone on the carbonate platform.

图 2.3.12 （摄影：钱迈平） Fig. 2.3.12 （Photograph by Qian Maiping）

这些叠层石呈次圆柱状，分枝强烈散开，生长方向多变，时而直立生长，时而匍匐生长，时而匍匐再转直立生长，基本层凸度、继承性及对称性多变。

The stromatolite columns are subcylindrical and markedly divergent branching, especially characteristic horizontal or subhorizontal to vertical appeared. Lamination changed in form, often deeply bulging or asymmetrical.

图 2.3.13 （摄影：钱迈平） Fig. 2.3.13 (Photograph by Qian Maiping)

叠层石柱体之间充填大量叠层石碎块和白云岩砾，都是叠层石生长期间堆积的，因此也称为同生角砾，显示其处于受强烈潮汐水流及风浪反复冲击的潮间带高能水动力环境。

The abundance of broken stromatolite fragments and dolostone gravels in interspaces between stromatolite columns showed these stromatolites were repeatedly impacted, even torn, by strong tidal currents and wind waves in a high hydrodynamic energy intertidal zone on the carbonate platform.

图2.3.14 （摄影：钱迈平） Fig. 2.3.14 (Photograph by Qian Maiping)

矿石山组中部浅灰色块状叠层石生物礁白云岩。叠层石生物礁由巨型的神农架大圆顶叠层石（*Megadomia shennongjiaensis*）构成，形成于碳酸盐台地潮下带水较深处。

Light grey massive stromatolite dolostones in the middle part of the Kuangshishan Formation. The bioherms were built up by giant domed stromatolites *Megadomia shennongjiaensis* which formed in a deeper subtidal zone on the carbonate platform.

图2.3.15 （摄影：钱迈平） Fig. 2.3.15 （Photograph by Qian Maiping）

这些叠层石的基本层呈巨大的穹状突起，单个叠层体大多宽5—6 m，高4—5 m。它们相互交错重叠，连绵延展，构成宏伟的巨型生物礁。

The huge arched laminae of these stromatolites are 5—6 m in width, 4—5 m high. They are alternately overlapped, extended and built up giant magnificent bioherms.

图 2.3.16 （摄影：钱迈平） Fig. 2.3.16 （Photograph by Qian Maiping）

这些巨型叠层石是神农架中元古代叠层石大堡礁群中，最令人难忘的一景。即使经历了10多亿年的岁月沧桑，今天依然巍然屹立。

These huge stromatolites are one of the most memorable scenes from the Mesoproterozoic stromatolite great barrier reefs. They still stand tall through the vicissitudes of more than 1 billion years.

图 2.3.17 （摄影：钱迈平）　　　　　　　　　Fig. 2.3.17　（Photograph by Qian Maiping）

矿石山组中部浅灰色块状叠层石生物礁白云岩。由大型的树桩圆柱叠层石(*Colonnella cormosa*)构成，形成于碳酸盐台地潮间带下部－潮下带上部。

Light grey massive stromatolite dolostones in the middle part of the Kuangshishan Formation. The bioherms were built up by big columnar stromatolites *Colonnella cormosa* formed in the low intertidal and shallow subtidal zone on the carbonate platform.

图 2.3.18 （摄影：钱迈平）　　　　　　　　　　Fig. 2.3.18　（Photograph by Qian Maiping）

这些叠层石个体呈不分枝的圆柱状，大多高度超过 1 m，相互紧密平行排列，间距不超过 50 mm。

These big non-branching cylindrical stromatolites are usually more than 1 m high, in vertically parallel arrangement, no more than 0.05 m apart.

图 2.3.19 （摄影：钱迈平） Fig. 2.3.19 （Photograph by Qian Maiping）

柱体横断面呈圆形，直径大多为 0.4—0.6 m。

Their columns are 0.4—0.6 m in diameter.

图 2.3.20 （摄影：钱迈平） Fig. 2.3.20 （Photograph by Qian Maiping）

矿石山组中部浅灰色块状叠层石生物礁白云岩。叠层石生物礁由大型的加尔加诺锥叠层石（*Conophyton garganicum*）构成，形成于碳酸盐台地潮下带上部。叠层石个体呈不分枝的锥柱状，横断面呈圆－次圆形，直径大多为 0.4－0.6 m，高度超过 1 m，相互紧密平行排列，间距不超过 50 mm，柱体之间充填叠层石碎块及白云岩角砾，显示其处于受强烈潮汐水流及风浪反复冲击的高能水动力环境。

Light grey massive stromatolite dolostones in the middle of the Kuangshishan Formation. The bioherm consisted of big conical stromatolites *Conophyton garganicum* formed in a shallow subtidal zone on the carbonate platform. These big non-branching conical columnar stromatolites are round or subround in transverse section, normally 0.4－0.6 m in diameter, more than 1 m high, in vertically parallel arrangement, no more than 5 cm apart. Numerous broken stromatolite fragments and dolostone gravels in interspaces between stromatolite columns showed these stromatolites were repeatedly impacted, even torn, by strong tidal currents and wind waves in a high hydrodynamic energy condition.

图 2.3.21 （摄影：钱迈平）　　　　　　　　　　Fig. 2.3.21 （Photograph by Qian Maiping）

矿石山组中部深灰色中厚层状自碎角砾白云岩。

Dark grey medium- to thick-bedded autoclastic brecciated dolostones in the middle part of the Kuangshishan Formation.

图 2.3.22 （摄影：钱迈平）　　　　　　　Fig. 2.3.22 （Photograph by Qian Maiping）

自碎角砾白云岩中网状方解石注射脉非常发育，但不延伸进入上覆岩层，显示在其成岩过程中或成岩后不久，即遭受强烈震动而碎裂。这些生成在硬岩层中的脉，通常密集成群向各个方向爆发延伸，将岩石切割成七巧板模式（自碎角砾），这是判别震积岩的重要标志。

Anastomosing-injection calcite dykes well developed in the autoclastic brecciated dolostones and are not extended into the overlying stratum. They are earthquake-induced structures formed during and soon after lithogenesis. These dykes generated in hard bedrock are generally in swarms, bursting out in all directions. Consequently, the cutting out the rocks produces a jigsaw-puzzle pattern （autoclastic breccias）. The geometries of the dyke swarms injected in all directions are good criteria to classify these dykes among seismites.

图 2.3.23 （摄影：钱迈平）　　　　　　　　　　Fig. 2.3.23 　（Photograph by Qian Maiping）

矿石山组中部浅灰色中厚层状叠层石生物礁白云岩。生物礁由层锥叠层石（*Straticonophyton*）构成，形成于碳酸盐台地潮间带中—下部。

Light grey medium- to thick-bedded stromatolite dolostones in the middle part of the Kuangshishan Formation. The bioherms were built up by conical-stratiform stromatolites *Straticonophyton* formed in the middle-low intertidal zone on the carbonate platform.

图 2.3.24 （摄影：钱迈平） Fig. 2.3.24 (Photograph by Qian Maiping)

其叠层石基本层混乱变形破碎，是因地震而导致沉积物液化造成的结果。

The laminae of these stromatolites are disorganised and intruded by the earthquake-induced liquefied sediments.

图 2.3.25 （摄影：钱迈平）　　　　　Fig. 2.3.25　（Photograph by Qian Maiping）

矿石山组中部浅灰色中厚层状叠层石生物礁白云岩。生物礁由锥叠层石（*Conophyton*）构成，形成于碳酸盐台地潮下带上部。

Light grey medium- to thick-bedded stromatolite dolostones in the middle part of the Kuangshishan Formation. The bioherms consisted of conical stromatolites *Conophyton* formed in the shallow subtidal zone on the carbonate platform.

图 2.3.26 （摄影：钱迈平） Fig. 2.3.26 (Photograph by Qian Maiping)

矿石山组中部深灰色中厚层状自碎角砾白云岩。网状方解石注射脉非常发育，是明显的震碎构造。

Dark grey medium- to thick-bedded autoclastic brecciated dolostones in the middle part of the Kuangshishan Formation. Developed anastomosing-injection calcite dykes in the dolostones are clearly quake-induced structures.

图 2.3.27 （摄影：钱迈平）　　　　　　　　　Fig. 2.3.27　（Photograph by Qian Maiping）

矿石山组中部青灰色中厚层状叠层石生物礁白云岩。生物礁由层叠层石（*Stratifera*）构成，形成于碳酸盐台地潮坪中－下部，其叠层石基本层混乱变形，上拱破裂，是因地震而导致沉积物液化并高压上涌造成的结果。

Cinerous medium- to thick-bedded stromatolite dolostones in the middle part of the Kuangshishan Formation. The biostromes were built up by stratiform stromatolites *Stratifera* formed in the middle-low intertidal zone on the carbonate platform. The laminae of these stromatolites are disorganized, ripped and intruded by the earthquake-induced overpressure liquefied sediments.

图 2.3.28 （摄影：钱迈平）　　　　　　　　　Fig. 2.3.28 （Photograph by Qian Maiping）

矿石山组上部青灰色厚层－块状自碎角砾叠层石生物礁白云岩。生物礁由地窖印卓尔叠层石（*Inzeria intia*）构成，叠层体呈棕黄色，显示其含有氧化铁。岩层网状方解石注射脉非常发育，但不延伸到上覆岩层，显示在其成岩过程中或成岩后不久曾遭受强烈震动而破碎。

Cinerous thick-bedded to massive autoclastic breccia stromatolite dolostones in the upper part of the Kuangshishan Formation. The biostromes were built up by tan columnar stromatolites *Inzeria intia* showed the ferric oxide component had been accumulated in these stromatlites. The dolostones with developed anastomosing-injection calcite dykes which are not extended into the overlying stratum suggested that the dolostones were hit by some of strong quakes and broken during and soon after the rock formation.

图2.3.29 （摄影：钱迈平）　　　　　　　　　　　　Fig. 2.3.29　（Photograph by Qian Maiping）

　　叠层体呈次圆柱-块茎状、芽状分枝及连层发育，形成于碳酸盐台地潮坪中部。

These stromatolites exhibit subcylindrical, tuberous and budded branching columns with connecting bridges. They formed on a middle tidal flat of the carbonate platform.

图 2.3.30 （摄影：钱迈平）　　　　　　　　Fig. 2.3.30　（Photograph by Qian Maiping）

从叠层体横断面看，这些叠层石为适应所处的潮坪环境，相邻叠层体相互相嵌生长，以充分利用光照面积，进行光合作用；并相互紧密依靠，组成结构坚固的生物礁，以抗击潮汐水流的反复冲击。

In transverse section, stromatolite columns interdigitated each other indicate the microbial mats of these stromatolites tended to make full use of the illumination area for photosynthesis and built up firm bioherms against the constant assault from tidal waves.

图 2.3.31 （摄影：钱迈平）　　　　　　　　Fig. 2.3.31　（Photograph by Qian Maiping）

矿石山组上部浅灰色块状叠层石生物礁白云岩。生物礁由巨型神农架大圆顶叠层石（*Megadomia shennongjiaensis*）构成，形成于碳酸盐台地潮下带水较深处。

Light grey massive stromatolite dolostones in the upper part of the Kuangshishan Formation. The bioherms were built up by giant domed stromatolites *Megadomia shennongjiaensis* formed in a deeper subtidal zone on the carbonate platform.

图 2.3.32 （摄影：钱迈平）

Fig. 2.3.32 (Photograph by Qian Maiping)

叠层体呈巨大的圆丘状，大多宽超过 6 m，高超过 5 m。

The huge arched laminae of these stromatolites are mostly larger than 6 m in width, more than 5 m high.

图 2.3.33
(摄影:钱迈平)

Fig. 2.3.33
(Photograph by Qian Maiping)

它们相互重叠或交错生长,构成巨大的生物礁。

These domes are alternately overlapped and built up gigantic bioherms.

图 2.3.34
(摄影:钱迈平)

Fig. 2.3.34
(Photograph by Qian Maiping)

大圆丘基底中心包含若干紧密并列生长的小丘状或短柱状叠层石。

The base of a giant domed stromatolite was formed by a cluster of smaller column stromatolites.

图 2.3.35 （摄影：钱迈平） Fig. 2.3.35 （Photograph by Qian Maiping）

  矿石山组上部浅灰色中厚层状自碎角砾白云岩。黑色网状燧石注射脉密集分布，不延伸进入上覆岩层，是在成岩期间或成岩后不久发生的强烈震动造成的碎裂构造。

Light grey medium- to thick-bedded autoclastic breccia dolostones in the upper part of the Kuangshishan Formation. The dolostones with well developed blake anastomosing-injection chert dykes which are not extended into the overlying stratum suggested that the dolostones were broken by some of strong quakes during and soon after lithogenesis.

图 2.3.36
(摄影:钱迈平)

Fig. 2.3.36
(Photograph by Qian Maiping)

矿石山组上部浅灰中薄层状破碎白云岩。方解石注射脉非常发育,并对应同生破碎构造,蜿蜒曲折贯穿切割岩层,记录了当时地震产生的高压注射作用。

Light grey medium- to thin-bedded broken dolostones in the upper part of the Kuangshishan Formation. Calcite injection dykes well developed and correspond to syndiagenetic broken structures, cutting squiggly through the beds, and recorded the higher rate of compaction resulted by the earthquake.

图 2.3.37 （摄影：钱迈平）　　　　　　　　　Fig. 2.3.37　（Photograph by Qian Maiping）

矿石山组上部浅灰色薄层状白云岩。交错层理发育，显示其处于强水流环境。

Light grey thin-bedded dolostones in the upper part of the Kuangshishan Formation. Developed well cross-beddings indicate a carbonate platform which were affected by strong water currents.

图 2.3.38 （摄影：钱迈平）

Fig. 2.3.38 （Photograph by Qian Maiping）

矿石山组顶部浅灰色中薄层状自碎角砾白云岩。

Light grey medium- to thin-bedded autoclastic breccia dolostones in the top of the Kuangshishan Formation.

矿石组地层沉积区由早先碳酸盐台地外较深水区逐渐向台地潮坪浅水转变。随着海水的变浅,沉积形成的岩石依次由灰黑色的页岩及铁质白云岩向浅灰色叠层石生物礁白云岩转变。叠层石类型包括:巨型的神农架大圆顶叠层石(*Megadomia shennongjiaensis*)、大型的加尔加诺锥叠层石(*Conophyton garganicum*)、地窖印卓尔叠层石(*Inzeria intia*)和波层叠层石(*Stratifera undata*),以潮下带及潮间带叠层石为主。因临近地壳构造活动带,受火山及地震作用,含铁热液沉积及地震沉积岩发育。

The Kuangshishan Formation was initially sedimented in a deeper water argillaceous depositional setting, later in a shallow water carbonate depositional setting as the water became gradually shallower. The sedimentary rocks turned from grey black shales, ferruginous dolostones to light grey stromatolite dolostones. There are developed stromatolite bioherms and biostromes in this formation, in which stromatolites exhibit a variety of forms and structures, or morphologies, including giant domal *Megadomia shennongjiaensis*, big conical *Conophyton garganicum*, big columnar *Colonnella cormosa*, middle columnar *Inzeria intia* and stratiform *Stratifera undata*. They were mainly formed in subtidal and intertidal zones. Developed ferruginous dolostones and seismites suggested that it had been located near an active submarine volcano and earthquake zone.

2.4　台子组

　　台子组厚744 m,根据岩性可分上、下两部分:下部以砂岩、粉砂岩、角砾岩及各种白云岩为特征,叠层石生物礁白云岩发育;上部以碳质泥岩及页岩为主。

　　沉积构造以水平层理为主,条带状构造发育。此外还有波状层理、斜层理、波痕、瘤状、豆状、鲕粒以及包括自碎角砾和网状注射脉等地震沉积构造。

　　与下伏地层矿石山组顶部浅灰色中薄层状自碎角砾白云岩呈假整合接触。

2.4　TAIZI FORMATION

　　Taizi Formation is 744 m thick and can be divided roughly into two lithologic parts. The lower Taizi part is consisted of siltstones, sandstones, breccias and various dolostones, including developed stromatolite dolostones. The upper Taizi part is mainly composed of carbargilites and shales.

　　The sedimentary structures in this formation are dominated by banded horizontal bedding, along with current bedding, cross bedding, ripple marks, nodules, pisolites, oolites and earthquake-induced autoclastic breccias, anastomosing-injection dykes, liquefied disorganizations and intrusions etc.

　　This formation is in disconformable contact with the underlying light gray midium- to thin- bedded autoclastic breccias in the top of the Kuangshishan Formation.

图 2.4.1 （摄影：钱迈平）　　　　　　Fig. 2.4.1 （Photograph by Qian Maiping）

　　台子组下部黑灰色条带状含炭质白云质粉砂岩。是在浪基面之下低能水动力且缺氧的还原环境中沉积形成。

Grey black dolomitic brecciated siltstones in the lower part of the Taizi Formation. They were deposited in a deeper water, low energy hydrodynamic and anoxic reductive environment below the wave base.

图 2.4.2 （摄影：钱迈平） Fig. 2.4.2 （Photograph by Qian Maiping）

网状注入脉发育。这是一种震碎构造，是这套岩石成岩期间或成岩后不久受强烈震动后破碎，破碎角砾间注入液化方解石而形成的。

Anastomosing-injection dykes developed in the dolomitic brecciated siltstones are shredded structures resulted from a violent quake during or short after the rock formation.

图 2.4.3 （摄影：钱迈平） Fig. 2.4.3 （Photograph by Qian Maiping）

台子组下部浅灰色中厚层状纹层白云岩。沉积于碳酸盐台地的浪基面以下较平静的低能水动力环境。

Light grey medium- to thick-bedded laminar dolostones in the lower part of the Taizi Formation. They deposited in a calm, low energy hydrodynamic environment on the carbonate platform below the wave base.

图 2.4.4 （摄影：钱迈平） Fig. 2.4.4 （Photograph by Qian Maiping）

台子组下部深灰色中厚层状纹层白云质粉砂岩。纹层发育，是在碳酸盐台地浪基面之下较平静的低能水动力环境中沉积形成。

Dark grey medium- to thick-bedded laminar dolomitic siltstones in the lower part of the Taizi Formation. The laminated beds formed in calm and low energy hydrodynamic environment on the carbonate platform below the wave base.

图 2.4.5 （摄影：钱迈平）　　　　　　　　　　Fig. 2.4.5　（Photograph by Qian Maiping）

台子组下部灰色中厚层状白云质角砾岩。是碳酸盐台地斜坡重力流堆积。

Grey medium- to thick-bedded dolomitic breccias in the lower part of the Taizi Formation. It was a sediment of gravity flow deposited on a carbonate platform slope.

图 2.4.6 （摄影：钱迈平）　　　　　　　　　Fig. 2.4.6 （Photograph by Qian Maiping）

台子组下部浅灰色中厚－中薄层状纹层泥质白云岩。是在碳酸盐台地浪基面以下浅水、较平静的低能水动力环境中沉积形成。

Light grey medium- to thick-bedded or medium- to thin-bedded laminar dolostones in the lower part of the Taizi Formation. It was deposited in a calm and low energy hydrodynamic environment on the carbonate platform below the wave base.

图 2.4.7 （摄影：钱迈平）　　　　　　　　　　　Fig. 2.4.7　（Photograph by Qian Maiping）

　　细密的深色－浅色交替的泥晶纹层夹较厚的薄层，显示其碳酸盐沉积在较平静的环境中随气候、季节或海水化学成分的周期变化而变化。

Alternation of dark and light colors micritic laminae with thin-bedded interbeds suggest that they be periodically deposited in a calm environment with climatic, seasonal or chemical composition of seawater changes cyclically.

图 2.4.8　（摄影：钱迈平）　　　　　　　　　Fig. 2.4.8　（Photograph by Qian Maiping）

台子组下部深灰色中薄－薄层状纹层白云质粉砂岩。是在碳酸盐台地外围浪基面以下较深水、平静的低能水动力环境中沉积形成。

Dark grey medium- to thin-bedded and thin-bedded laminar dolomitic siltstones in the lower part of the Taizi Formation. They were deposited in a low energy hydrodynamic environment on the periphery of the carbonate platform below the wave base.

图2.4.9 （摄影：钱迈平）　　　　　　　　　　Fig. 2.4.9　（Photograph by Qian Maiping）

台子组下部灰色中厚层状白云质角砾岩。是碳酸盐台地斜坡重力流堆积。

Grey medium- to thick-bedded dolomitic breccias in the lower part of the Taizi Formation. It was a sediment of gravity flow deposited on a carbonate platform slope.

图 2.4.10 （摄影：钱迈平） Fig. 2.4.10 （Photograph by Qian Maiping）

台子组下部深灰色中薄－中厚层状纹层白云质粉砂岩。是在碳酸盐台地外围浪基面以下较深水、平静的低能水动力环境中沉积形成。

Dark grey medium- to thin-bedded and medium- to thick-bedded laminar dolomitic siltstones in the lower part of the Taizi Formation. They were deposited in a calm deep water and lower hydrodynamic environment on the periphery of the carbonate platform below the wave base.

图 2.4.11 （摄影：钱迈平） Fig. 2.4.11 （Photograph by Qian Maiping）

台子组下部深灰紫色厚层块状白云质角砾岩。砾石及基质以白云质粉砂岩为主，砾石分选和磨圆度差，大小不一，分布无序，是碳酸盐台地边缘斜坡重力流快速堆积。

Dark grey purple massive dolomitic breccias in the lower part of the Taizi Formation. It made primarily of clasters and matrix of the dolomitic siltstones. These angular to subangular, unsized and randomly oriented clasters show a sediment of gravity flow rapidly accumulated on the carbonate platform slope.

图 2.4.12 （摄影：钱迈平）　　　　　　　　Fig. 2.4.12　（Photograph by Qian Maiping）

台子组下部浅灰色块状叠层石礁白云岩。由巨型的神农架大圆顶叠层石（*Megadomia shennongjiaensis*）构成，形成于碳酸盐台地潮下带水较深处。

Light grey massive stromatolite dolostones in the lower part of the Taizi Formation. The bioherms consisted of giant domed stromatolites *Megadomia shennongjiaensis* formed in a deeper subtidal zone on the carbonate platform.

图 2.4.13
（摄影：钱迈平）

Fig. 2.4.13
(Photograph by Qian Maiping)

受火山热液作用，叠层石礁白云岩轻度硅化。

The stromatolite dolostones were partly silicified by volcanic hydrothermal fluids.

图 2.4.14
（摄影：钱迈平）

Fig. 2.4.14
(Photograph by Qian Maiping)

因叠层石基本层的富有机质暗层与贫有机质的亮层硅化上的差异，岩石表面风化后凸凹鲜明。

Very distinct convex-concave stromatolitic laminae on the weathered surfaces of these bioherms show the difference in silica content between the silicified light and dark color laminae.

图 2.4.15 （摄影：钱迈平）　　　　　　　　　Fig. 2.4.15　（Photograph by Qian Maiping）

叠层石基本层波状起伏不定，显示当时的微生物席表面分布许多大小突起，体现了光合作用微生物的趋光特性。

Undulated irregularly stromatolitic laminae suggested there be a number of big and little phototatic microbial mat uplifts.

图 2.4.16 （摄影：钱迈平） Fig. 2.4.16 （Photograph by Qian Maiping）

台子组下部浅灰色块状叠层石生物礁白云岩。由大型的加尔加诺锥叠层石（*Conophyton garganicum*）构成，这种叠层石通常形成于潮下带较深海底。

Light grey massive stromatolite dolostones in the lower part of the Taizi Formation. The bioherms consisted of conical stromatolites *Conophyton garganicum* formed in a deeper subtidal zone on the carbonate platform.

图 2.4.17
(摄影:钱迈平)

Fig.2.4.17
(Photograph by Qian Maiping)

这些大型叠层石个体呈不分枝的圆锥柱状,大多高度超过 1 m,直径 0.4—0.6 m,相互紧密平行排列,间距不超过 50 mm,充填白云岩角砾,显示其形成时深度仍处于浪基面以上。

These big non-branching conical stromatolites are round or subround in transverse section, usually more than 1 m high, 0.4—0.6 m in diameter, in vertically parallel arrangement, no more than 5 cm apart. Numerous broken dolostone gravels in interspaces between stromatolite columns showed these stromatolites were in a very turbulent environment above the wave base.

图 2.4.18 （摄影：钱迈平） Fig. 2.4.18 （Photograph by Qian Maiping）

台子组下部浅灰色块状叠层石礁白云岩。由巨型的神农架大圆顶叠层石（*Megadomia shennongjiaensis*）构成，形成于碳酸盐台地潮下带较深海底。

Light grey massive stromatolite dolostones in the lower part of the Taizi Formation. The bioherms consisted of giant domed stromatolites *Megadomia shennongjiaensis* formed in a deeper subtidal zone on the carbonate platform.

图 2.4.19 （摄影：钱迈平）　　　　　　　Fig. 2.4.19　（Photograph by Qian Maiping）

台子组下部浅灰色块状叠层石生物礁白云岩。由大型的加尔加诺锥叠层石（*Conophyton garganicum*）构成，这种叠层石通常形成于潮下带较深海底。

Light grey massive stromatolite dolostones in the lower part of the Taizi Formation. The bioherm consisted of big conical stromatolites *Conophyton garganicum* formed in a deeper subtidal zone on the carbonate platform.

图 2.4.20
(摄影:钱迈平)

Fig. 2.4.20
(Photograph by Qian Maiping)

这些大型叠层石个体呈不分枝的圆锥柱状,大多高度超过 1 m,直径 0.4—0.6 m,相互紧密平行排列,间距不超过 50 mm,柱体之间充填白云岩角砾,显示其处于动荡环境,形成深度在浪基面以上。

These big non-branching conical columnar stromatolites are round or subround in transverse section, normally more than 1 m high, 0.4—0.6 m in diameter, in vertically parallel arrangement, no more than 5 cm apart. Numerous broken dolostone gravels in interspaces between stromatolite columns showed these stromatolites were in a very turbulent environment above the wave base.

图 2.4.21 （摄影：钱迈平）

Fig. 2.4.21 (Photograph by Qian Maiping)

加尔加诺锥叠层石（*Conophyton garganicum*）基本层在轴部明显突起，是当时光合作用微生物在光照较弱的潮下带较深水下具强烈趋光行为的表现。

The contorted conical laminae of *Conophyton garganicum* in their crestal parts show microbial mat uplifts of a strong phototatic response in a low light environment in the deeper subtidal zone.

图 2.4.22 （摄影：钱迈平）　　　　　　　　　　Fig. 2.4.22　（Photograph by Qian Maiping）

台子组下部浅灰色中厚层状叠层石生物礁白云岩。由横断面呈圆形-次圆形的树桩圆柱叠层石(*Colonnella cormosa*)构成,形成于潮间带中下部。

Light grey medium - to thick-bedded stromatolite dolostones in the lower part of the Taizi Formation. The bioherms were built up by cylindrical to subcylindrical columnar stromatolites *Colonnella cormosa* formed in a mid-lower intertidal zone on the carbonate platform.

图 2.4.23 （摄影：钱迈平）　　　　　　　　　　　Fig. 2.4.23　（Photograph by Qian Maiping）

台子组下部浅灰色块状叠层石生物礁白云岩。主要由大型的加尔加诺锥叠层石（*Conophyton garganicum*）构成，这种叠层石通常形成于碳酸盐台地潮坪下部较深海底。其纵横交错的黑色燧石细脉，是后期火山热液充填岩石裂隙而形成。

Light grey massive stromatolite dolostones in the lower part of the Taizi Formation. The bioherm consisted of big conical stromatolites *Conophyton garganicum* formed in a deeper subtidal zone on the carbonate platform. Crisscrossing black chert veinlets in the dolostones resulted from volcanic hydrothermal filling fissures in the rocks.

图 2.4.24 （摄影：钱迈平） Fig. 2.4.24 （Photograph by Qian Maiping）

　　这些大型叠层石个体呈不分枝的圆锥柱状，大多高度超过 1 m，直径为 0.4—0.6 m，相互紧密平行排列，间距不超过 50 mm，充填白云岩角砾，显示其形成时深度处于浪基面以上很动荡的环境。

These big non-branching conical columnar stromatolites are generally more than 1 m in high and 0.4—0.6 m in diameter. Stromatolitic columns in vertically parallel arrangement, no more than 0.5 m apart, and a lot of broken dolostone gravels in interspaces between the columns showed these stromatolites were in a very turbulent environment above the wave base.

图2.4.25 （摄影：钱迈平） Fig. 2.4.25 (Photograph by Qian Maiping)

台子组下部浅灰色中厚－中薄层状白云质粉砂质泥岩。形成于碳酸盐台地较平静的潟湖环境。

Light grey medium- to thick-bedded and medium- to thin-bedded laminar dolostones in the lower part of the Taizi Formation. They were formed in a calm lagoon environment on the carbonate platform.

图 2.4.26 （摄影：钱迈平）　　　　　　　　　Fig. 2.4.26 （Photograph by Qian Maiping）

台子组下部浅灰色中厚层状瘤状白云岩。风化面呈浅紫灰色，瘤块风化掉后，留下密集的孔洞。其原岩形成于碳酸盐台地较平静的潟湖环境。

Light grey medium- to thick-bedded nodular dolostones in the lower part of the Taizi Formation. Weathered surfaces of the nodular dolostones are light purple gray in color with densely distributed holes which are vestiges of nodules had been weathered out of their host rock formed in a calm lagoon environment on the carbonate platform.

图 2.4.27 （摄影：钱迈平）　　　　　　　　　　Fig. 2.4.27 （Photograph by Qian Maiping）

其瘤块约占 70%，基质约占 30%。瘤块色浅，多为较纯的白云岩，形状不规则，紧密镶嵌分布，长轴顺层排列；基质色深，为白云质粉砂质泥岩。

The nodules, about 70% of the nodular dolostones, are lighter in color, more pure dolomites, very irregular shape, commonly elongate and oriented parallel to the bedding. The matrix, about 30 % of the rock, is a dolomitic silty mudstones and darker in color.

图 2.4.28 （摄影：钱迈平） Fig. 2.4.28 (Photograph by Qian Maiping)

层面上可见瘤块叠覆，无滚动迹象。

The nodules seen in a transverse section of the nodular dolostones are overlapped each other and no-rolled.

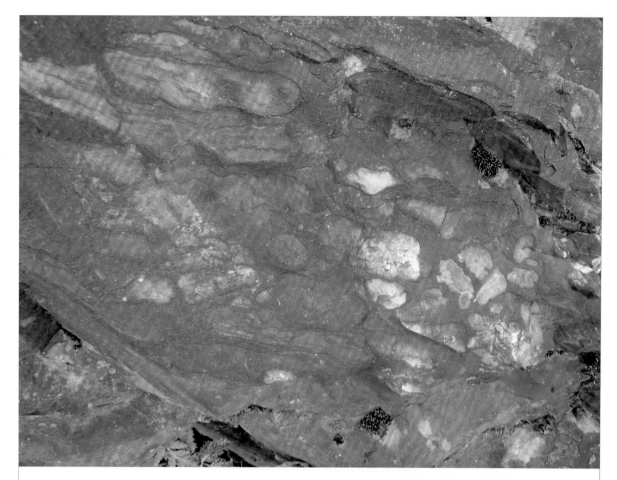

图 2.4.29 （摄影：钱迈平）　　　　　　　　　　Fig. 2.4.29　(Photograph by Qian Maiping)

纵剖面上可见瘤块镶嵌分布,显示这种瘤状白云岩原先是薄层白云岩与页岩交替沉积于碳酸盐台地的近岸潟湖浅水环境,在成岩过程中,受压实作用,发生横向运动,半固结的白云岩塑性差,被压裂或拉断成不规则瘤块；而同时期的页岩塑性好,被压薄并挤压进瘤块之间的空隙,最终形成瘤状白云岩。

The nodules seen in a vertical section of the nodular dolostones are irregular dehisced fragments. During the sedimentation of the nodular dolostones, this area located in a lagoon environment with shallow water and offshore on the carbonate platform. Dolomite and micrite beds formed by alternational sedimentation formed the nodular dolostones under the pressure, in which the semisolided mass of dolomite fractured and formed the nodules under the pressure. The soft micrite sediments removed upward and formed the flame and flow stractures.

图2.4.30 （摄影：钱迈平） Fig. 2.4.30 （Photograph by Qian Maiping）

台子组上部灰黑色薄层状泥岩及页岩。细腻的泥质结构和水平层理沉积构造，反映其形成于碳酸盐台地外较深水下的平静环境。

Grey black thin-bedded mudstones and shales in the upper part of the Taizi Formation. Their fine-grained texture and horizontal bedding structure indicate a calm and deeper water environment outside of the carbonate platform.

图 2.4.31
(摄影:钱迈平)

Fig. 2.4.31
(Photograph by Qian Maiping)

灰黑色显示其沉积于水体停滞的缺氧海底。

Their grey black color explains a stagnant water body in the anoxic environment in which they were formed.

图 2.4.32
(摄影:钱迈平)

Fig. 2.4.32
(Photograph by Qian Maiping)

台子组上部辉绿岩侵入体。由地幔的玄武质岩浆沿裂隙侵入到地壳岩石中冷却形成。

A diabase intrusive body in the upper part of the Taizi Formation. It was one of the basic dykes mainly composed of basaltic magma from the mantle filled into the rock according to the tensile fractures.

图 2.4.33 （摄影：钱迈平） Fig. 2.4.33 （Photograph by Qian Maiping）

辉绿岩是一种铁镁质全结晶潜火山岩，形成于超过 2 km 深度的中—浅部地壳内。通常呈中—细粒结构，肉眼可见板条状斜长石晶体分布在更细粒的辉石矿物基质里。

Diabase is a mafic holocrystalline subvolcanic rock that is emplaced at medium to shallow depths (more than 2 km) within the crust. It normally has a medium-to fine-grained, but visible texture of white euhedral lath-shaped plagioclase crystals set in a finer matrix of black clinopyroxene.

图 2.4.34 （摄影：钱迈平）　　　　　　　　Fig. 2.4.34　（Photograph by Qian Maiping）

辉绿岩致密坚硬，主要由斜长石、单斜辉石以及磁铁矿－钛铁矿组成，伴随和蚀变的矿物包括：石英、角闪石、黑云母、磷灰石、锆石、磁黄铁矿、黄铜矿、蛇纹石、绿泥石、方解石和叶腊石。

Diabase is a dense, hard rock composed mostly of plagioclase, clinopyroxene and magnetite-ilmenite. Accessory and alteration minerals include quartz, hornblende, biotite, apatite, zircon, pyrrhotite, chalcopyrite, serpentine, chlorite, calcite and pyrauxite.

图 2.4.35 （摄影：钱迈平）　　　　　　　　　　Fig. 2.4.35　（Photograph by Qian Maiping）

　　台子组上部灰黑色中薄层状硅质泥岩及页岩。水平层理显示其形成于碳酸盐台地外较深水下的平静环境。

Grey black medium- to thin-bedded siliceous mudstones and shales in the upper part of the Taizi Formation. Their horizontal bedding shows a calm and deeper water environment outside of the carbonate platform.

图 2.4.36
（摄影：钱迈平）

Fig. 2.4.36
(Photograph by Qian Maiping)

灰黑色显示其沉积于水体停滞的缺氧海底。

Their grey black color explains a stagnant anoxic environment in which they were formed.

图 2.4.37
（摄影：钱迈平）

Fig. 2.4.37
(Photograph by Qian Maiping)

浅灰色条带构造非常发育，显示沉积物成分呈周期性变化。

Developed light grey banded structures indicate the composition of sediments varied cyclically.

图 2.4.38 （摄影：钱迈平） Fig. 2.4.38 (Photograph by Qian Maiping)

台子组上部灰黑色中厚－中薄层状泥岩与薄层状纹层泥岩互层。显示其形成于碳酸盐台地边缘外较深水下的平静环境，沉积量随季节周期变化。

Cyclically alternating grey black medium- to thick-bedded, medium- to thin-bedded mudstones and thin-bedded laminar mudstones in the upper part of the Taizi Formation. They show the amount of sediment varied seasonally in a calm and deeper water environment outside of the carbonate platform.

图 2.4.39 （摄影：钱迈平）　　　　　　　　　Fig. 2.4.39 （Photograph by Qian Maiping）

　　当时这里的气候特点是雨季－旱季交替。雨季泥沙入海量大，沉积速率快，形成中厚－中薄层状泥岩；而旱季泥沙入海量小，沉积速率慢，形成薄层状纹层泥岩。

Since the amount of sediment varied seasonally in a tropical climate, it was likely that the medium- to thick-bedded, medium- to thin-bedded mudstones were formed by high amount of sediment during the rainy season, while the thin-bedded laminar mudstones were formed by lower amount of sediment during the dry season.

图 2.4.40 （摄影：钱迈平） Fig. 2.4.40 （Photograph by Qian Maiping）

台子组上部黑色中厚－中薄层状泥岩与薄层状泥岩互层。沉积于碳酸盐台地外更深水下缺氧的还原环境，富含硫化氢（H_2S）、黄铁矿（FeS_2）和碳，呈现黑色。其风化面因黄铁矿被氧化，呈现暗红色。

Alternating black medium- to thick-bedded and medium- to thin-bedded mudstones in the upper part of the Taizi Formation. They were deposited in a deeper water environment outside of the carbonate platform. These black mudstones, which formed in an anoxic reducing condition, contained hydrogen sulfide (H_2S) and pyrite (FeS_2) along with carbon produced the black coloration. The weathered surfaces of these rocks appear black red stained by pyrite oxidation.

图 2.4.41 （摄影：钱迈平） Fig. 2.4.41 (Photograph by Qian Maiping)

这套黑色泥岩含钼，沿层理及裂隙可见灰白色三氧化钼（MoO_3）结晶粉末。钼在常温下很稳定，当温度升高超过400℃时，与氧气发生轻度反应，稍有氧化；温度超过600℃时，迅速与氧发生反应，氧化生成灰白色的三氧化钼；超过700℃时，与水蒸气发生强氧化反应，生成紫色的二氧化钼（MoO_2）。因此，这套黑色泥岩成岩后曾被附近侵入的岩浆加热到超过600℃，但不到700℃。

Grey white molybdenum trioxide (MoO_3) powder occurred along layers and fractures in the black mudstones revealed that these rocks contain molybdenum. Molybdenum chemical properties in the air at room temperature are stable. When the temperature is higher than 400℃, there is slight oxidation. When the temperature is higher than 600℃, the metal will rapidly oxidize to gray white molybdenum trioxide. When the temperature is higher than 700℃, the water vapor will oxidize Mo into violet molybdenum dioxide (MoO_2). It shows the black mudstones had been heated to above 600℃ and below 700℃ by magma intrusions.

图 2.4.42 （摄影：钱迈平） Fig. 2.4.42 （Photograph by Qian Maiping）

野马河公路沿线出露的地层：台子组上部黑色中薄层状页岩。沉积于碳酸盐台地外较深水下缺氧的还原环境。

Along Yemahe highway: Black medium- to thin-bedded shales in the upper part of the Taizi Formation. They were deposited in a deeper water anoxic reducing condition outside of the carbonate platform.

图 2.4.43 （摄影：钱迈平）　　　　Fig. 2.4.43 （Photograph by Qian Maiping）

这套黑色页岩富含钒，钒沿破碎构造及裂隙被氧化后，形成一坨坨橙黄色钒氧化物——五氧化二钒（V_2O_5）。钒在常温中不会被氧化，而在高温中很容易与氧和氮发生反应，氧化成棕黑色的三氧化二钒（V_2O_3）、深蓝色的二氧化钒（VO_2），最终成为橙黄色的五氧化二钒。可见，这套黑色页岩成岩后曾受过附近侵入的岩浆高温作用。

Orange vanadium pentoxide (V_2O_5) solids occurred along fractures around in the black shales indicate these rocks contain abundant vanadium. Vanadium does not be oxidized at room temperature, while it can easily react with oxygen and nitrogen at high temperature to form brownish black vanadium trioxide (V_2O_3), dark blue vanadium dioxide (VO_2), until orange vanadium pentoxide is reached. Therefore the black shales had been heated to high temperature by magma intrusions.

台子组地层沉积位置随时间的推移,逐渐由碳酸盐台地潮坪及边缘斜坡向边缘外盆地转变。处于台地潮坪及边缘斜坡时期,叠层石生物礁仍很发育,叠层石以潮间带中－下部至潮下带较深水的类型为主,如波层叠层石(*Stratifera undata*)、大－中型的树桩圆柱叠层石(*Colonnella cormosa*)、加尔加诺锥叠层石(*Conophyton garganicum*)和巨型的神农架大圆顶叠层石(*Megadomia shennongjiaensis*),并时有斜坡重力流形成的砾岩层。处于台地外盆地时期,海水已加深到光合作用无法进行的深度,叠层石随之消失。根据现代海洋学研究,近海因富含碎屑,通常在超过50 m水深处光合作用就难以进行;远洋因透明度较好,在水深200 m处仍可进行光合作用。此时,因处于风暴浪基面以下,水体较平静,与表层富氧海水几乎没有交流,有利于沉积有机质的厌氧微生物矿化作用,形成的黑色泥岩和页岩富含有机碳、硫化氢(H_2S)、黄铁矿(FeS_2)及其他硫化矿物。成岩后受辉绿岩体侵入高温作用,导致三氧化钼(MoO_3)、五氧化二钒(V_2O_5)及其他氧化矿物的形成。

The deposits of the Taizi Formation interpret the seawater became deep gradually, and it transformed slowly from shallow water tidal flats, logons and slopes of the carbonate platform into a deeper water basin out of the platform. Bioherms and biostromes were built up by developed mid-lower intertidal and subtidal stromatolites *Stratifera undata*, *Colonnella cormosa*, *Conophyton garganicum* and *Megadomia shennongjiaensis* on the carbonate platform, along with sediment of gravity flow deposits on a carbonate platform slope. As depth continually increased the intensity of light rapidly dissipated. Such a minuscule amount of light penetrates beyond a depth of about 50 m in coastal waters or about 200 m in open ocean that photosynthesis is no longer possible, and stromatolites disappeared. In a calm anoxic environment below the storm wave base the black mudstones and shales were formed. They contain abundant organic carbon, hydrogen sulfide (H_2S), pyritic (FeS_2) and other sulfide minerals due to anaerobic mineralization of marine sediment organic matter. The molybdenum trioxide (MoO_3), vanadium pentoxide (V_2O_5) and other oxide minerals were formed when the black mudstones and shales were heated by the diabase intrusions.

2.5 野马河组

野马河组厚 1 369 m,以各种白云岩为特征,上部出现叠层石生物礁白云岩。

沉积构造包括水平层理、斜层理、条纹、搅动构造,有鲕粒及同生碎屑。

与下伏的台子组顶部黑色页岩整合接触。

2.5 YEMAHE FORMATION

Yemahe Formation is 1,369 m thick and consisted basically of various dolostones including several stromatolite biostrome-bioherm dolostones in the upper part of this formation.

The sedimentary structures in this formation includes horizontal bedding, cross bedding, banded and turbulent structures, along with oolites, breccias and detritals.

This formation is in conformable contact with the underlying black shale in the top of the Taizi Formaion.

图 2.5.1 （摄影：钱迈平）

Fig. 2.5.1 （Photograph by Qian Maiping）

野马河组底部的灰黑色中薄层状纹层白云岩，与下伏的台子组顶部黑色中薄层状页岩整合接触。

Grey black medium- to thin-bedded laminar dolostone in the bottom of the Yemahe Formation is in conformable contact with the underlying black shale in the top of the Taizi Formaion.

图2.5.2 （摄影：钱迈平）　　　　　　　　　Fig. 2.5.2　（Photograph by Qian Maiping）

野马河组下部灰黑色中薄层状纹层白云岩。显示其形成于热带气候下浪基面以下海底较稳定的低能水动力沉积环境，水体动荡小，沉积的碳酸盐灰泥颗粒很细。

Grey black medium- to thin-bedded laminar dolostones in the lower part of the Yemahe Formation. Their sedimentological character of fine-grained carbonate mud laminae indicates a relatively calm underwater environment below the wave base under a tropical condition.

图 2.5.3 （摄影：钱迈平）　　　　　　　　　　　　Fig. 2.5.3　（Photograph by Qian Maiping）

野马河组上部灰紫色中厚层状叠层石礁白云岩夹中薄层状斑脱岩。

风化面呈暗黄色，是其所含有的三价铁被氧化后显示的颜色。说明当时此处海水变浅，已处于发育叠层石生物礁的碳酸盐台地潮坪浅水区。因斑脱岩是火山喷发喷出的火山灰沉积形成凝灰岩后，经强风化而成的黏土岩，由此可见当时附近至少有一座活火山不时地喷发，沉积了多层火山灰。

Grey purple medium - to thick-bedded stromatolite dolostones with medium- to thin-bedded bentonites interbeds in the upper part of the Yemahe Formation.

The dark yellow weathered surface of the rock indicates high iron oxide content, and explains seawater became shallow, the environment was transformed into a tidal flat on the carbonate platform in which stromatolite bioherms developed well. Bentonites usually formed from weathering of volcanic ash-fall beds, so they suggested that there be at least an active volcano in near, and it often spewed a massive plume of volcanic ash into the air and fell here.

图 2.5.4 （摄影：钱迈平）　　　　　　　　　　Fig. 2.5.4 （Photograph by Qian Maiping）

生物礁由育卡贝加尔叠层石(*Baicalia unca*)构成,柱体呈块茎状或次圆柱状,基部收缩,继而膨胀,基本层生长继承性差,柱体以强烈散开式二分叉为主。显示当时处于碳酸盐台地潮间带,建造叠层石的微生物席受变化多端的潮汐、水深、水流、光照和营养条件影响,其生长面积和生长方向也随之变化,最终造就了叠层石的多变形态。

The stromatolite bioherms are consisted of *Baicalia unca*. Their tuberous or subcylindrical columns with marked Y-shaped diverging branches and variation in diameter were perhaps resulted from changeable areas and growth directions of the microbial mats in changeable tidal streams, water depths, ocean currents, sunlights and nutritional conditions in an intertidal zone on the carbonate platform.

图2.5.5 （摄影：钱迈平） Fig. 2.5.5 （Photograph by Qian Maiping）

野马河组上部灰紫色厚层－块状叠层石礁白云岩。也是由育卡贝加尔叠层石（*Baicalia unca*）构成。用其厚层状白云质凝灰岩夹层中的锆石，测试获得的铀铅同位素年龄约12.158亿年。

Grey purple thick-bedded to massive stromatolite dolostones in the upper part of the Yemahe Formation. The bioherms are consisted of tuberous or subcylindrical columnal stromatolites *Baicalia unca*. Their age is about 1,215.8 million years old. This dating is based on the evidence from uranium—lead radiometric dating of the mineral zircon from thick-bedded interbeds of dolomitic tuff in the bioherms.

野马河组地层自下而上，显示了海水逐渐变浅的过程。以中薄层状白云岩为特征，由较深水的缺氧还原环境沉积形成的灰黑色白云岩，逐渐转变为潮坪浅水干旱炎热气候强氧化环境沉积形成的紫红色白云岩，叠层石生物礁也随之出现。

The deposits of the Yemahe Formation descript the seawater became shallow gradually, and it transformed slowly the grey black dolostones formed in an anoxic reducing deep water environment into the purple red stromatolite dolostones formed in a strong oxidizing shallow water tidal flat environment in an arid and hot climate.

2.6　温水河组

温水河组厚 1 878 m,以白云岩为主,下部有三套玄武岩,上部出现叠层石生物礁白云岩。沉积构造以水平层理为主,条带状构造发育,还有搅动构造,伴有鲕粒、角砾、碎屑。与下伏地层野马河组顶部灰色白云岩整合接触。

2.6　WENSHUIHE FORMATION

Wenshuihe Formation is 1,878 m thick and consisted mainly of dolostones including three basalt units in lower part of this formation and several stromatolite biostrome-bioherm dolostones in the upper part of this formation.

The sedimentary structures in this formation are dominated by banded horizontal bedding, along with turbulent structures, oolites, breccias and detritals.

This formation is in conformable contact with the underlying gray dolostone in the top of the Yemahe Formaion.

图 2.6.1 （摄影：钱迈平）

Fig. 2.6.1 （Photograph by Qian Maiping）

图 2.6.2 （摄影：钱迈平）

Fig. 2.6.2 （Photograph by Qian Maiping）

　　玄武岩是一种火山喷出岩浆快速冷却形成的深色火山岩，主要成分是钙质斜长石、辉石及钛磁铁矿。玄武岩通常颜色发黑，其原因就是因为含有辉石类和磁铁矿类矿物。玄武岩显然是地球上最常见的火山岩，也常见于月球和太阳系其他岩质行星。为什么会这样常见？因为它是地壳最原始的组成部分，而几乎所有其他岩石类型都是由它演化出来的。

　　火山岩是由岩浆冷却结晶或固结而成的。在熔融的液态岩浆里，一个个原子可自由地四处运动。当岩浆冷却时，这些原子就失去能量并聚集起来形成固态的矿物晶体。晶体的生长只发生在岩浆尚未固结时，一旦完全固结，晶体即停止生长。所以，矿物晶体的大小取温水河组底部灰黑色块状玄武岩,直接覆盖在野马河组顶部灰色白云岩上。显示其当时处于海底火山喷出的岩浆漫溢范围内,距离火山口并不太远。

Grey black massive basalts in the bottom of the Wenshuihe Formation overlaid on the grey dolostones in the top of the Yemahe Formation shows they accumulated near a submarine volcanic vent location within a channel or reservoir of the lava overflow.

决于岩浆冷却的速度。如冷却很缓慢,矿物有充分的时间生长晶体,就会产生大个晶体。如冷却很快,晶体生长时间很少,晶体就很小。如冷却极快,一个个原子会即刻定格在原地,根本来不及形成晶体。温水河组下部的玄武岩正是这样,没有气孔或杏仁构造,是尚未喷出但在很接近地表处极其快速冷却形成的非晶质玻璃状岩石。

Basalt is a dark-colored extrusive volcanic rock formed from the rapid cooling of basaltic lava and composed usually of calcic plagioclase, augite and titaniferous magnetite. Black color is given to basalt by pyroxene and magnetite. Both of them contain iron and this is the reason why they are black. Basalt is clearly the most common volcanic rock on the Earth, and also common on the Moon and other rocky planets of the Solar System. What makes basalt so common? Basalt is the original constituent of the crust from which almost all other rock types have evolved.

Volcanic rocks are formed by the cooling and crystallization or solidification of lava. In liquid lava, individual atoms are free to move around. As a body of magma cools, these individual atoms lose energy and come together to form solid mineral crystals. Crystal growth only occurs when liquids are present. Once all liquids solidify, crystal growth stops. The size of crystals produced depends on the rate at which the lava cools. If cooling is very slow, minerals have lots of time to grow, and so large crystals are produced. If cooling is rapid, little time is available for growth, and so crystals will be small. If cooling is extremely rapid, individual atoms will be frozen in place and crystals may not form at all. The basalts in the lower part of the Wenshuihe Formation are amorphous substances without any vesicule formed from the extremely rapid cooling of basaltic lava at very near the surface of Earth.

图 2.6.3 （摄影：钱迈平）　　　　　　　　　　Fig.2.6.3　（Photograph by Qian Maiping）

玄武岩局部含大量白云岩角砾，是岩浆流动期间沿途裹挟的白云岩碎块。这些白云岩碎块的磨圆、分选度很差，显然是在火山喷发时，已固结成岩的白云岩层被震碎或炸碎，随即被流经的岩浆带走。

The basalt containing partially large amounts of angular dolomite fragments formed from the rapid cooling of basaltic lava carried fragments of the dolostones. These dolomite fragments are poorly rounded and sorted, and were held and carried by flowing lava shortly after the dolostones were shattered by a volcanic explosion.

图 2.6.4 （摄影：钱迈平） Fig. 2.6.4 （Photograph by Qian Maiping）

温水河组上部浅灰色块状叠层石生物礁白云岩。叠层石以地窖印卓尔叠层石（*Inzeria intia*）为主，叠层石柱体呈次圆柱－块茎状，相互靠紧，密集生长，柱体间空隙很小。

Light gray massive stromatolite dolostones in the upper part of the Wenshuihe Formation. The bioherms consisted of tuberous or subcylindrical columnar stromatolites *Inzeria intia*. Interspaces between the columns of these stromtolites are very narrow.

图2.6.5 （摄影：钱迈平） Fig. 2.6.5 （Photograph by Qian Maiping）

叠层石柱体横断面呈圆多边形,相互紧密镶嵌排列。显示这些叠层石生物礁形成于碳酸盐台地潮下带,光照不如潮间带那么充足,形成叠层石的微生物席为尽可能充分利用有限的光照面积进行光合作用,采取了紧密镶嵌排列的生长方式。

Columns are rounded polygons and interdigitated each other in transverse section. It indicates these stromatolites formed in a subtidal zone on the carbonate platform where the sunlight was less than that in an intertidal zone, therefore microbial mats interdigitated closely each other to make full use of the illumination area for photosynthesis .

温水河组地层沉积初期,位于喷发的海底火山岩浆漫溢范围,形成三套玄武岩。因主要处于碳酸盐台地潮下带,其沉积构造以较平静环境形成的水平层理为主,偶尔也出现反复振荡水体形成的鲕粒和短期强烈动荡水体形成的角砾。因水较深,叠层石在很有限的光照区域以密集镶嵌生长形成一些生物礁。

The Wenshuihe Formation was at first accumulated three basalt units within a channel or reservoir of the lava overflowed near a submarine volcanic vent. The location was mainly at an subtidal zone of the carbonate platform and therefore sedimentary structures are dominated by banded horizontal bedding, along with turbulent structures, oolites, breccias and detritals. Late, stromatolites developed and grew so closely together in a slightly deep water subtidal zone that their microbial mats interdigitated each other and formed some bioherms in a limited illumination area.

2.7 石槽河组

石槽河组厚1 656 m,根据岩性可分上、中、下三部分:

下部,以紫红色含铜纹层白云岩、紫红色泥质纹层白云岩－粉黄色白云质泥岩互层以及浅灰色含铜叠层石生物礁白云岩为特征。

中部,以深灰色叠层石生物礁白云岩、浅灰色硅质条带白云岩、灰紫－紫红色叠层石生物礁白云岩,以及紫红色含铜叠层石生物礁白云岩为特征。

上部,以杂色纹层白云岩与浅灰－肉红色纹层白云岩互层、黑灰色含硅质白云质粉砂岩、黑色含碳质白云质粉砂岩为特征。

沉积构造以水平层理为主,条带状构造发育,还有交错层理、波痕、泥裂、鲕粒、重力滑动、断裂、注射脉、同生碎屑及角砾。

2.7 SHICAOHE FORMATION

Shicaohe Formation is 1,656 m thick and can be divided into three units depending on their lithology: lower, middle and upper parts.

The lower part characterized by purple-red copper-bearing laminated dolostone, alternating purple-red muddy laminated dolostone and light pink-yellow dolomitic marlstone, as well as light grey copper-bearing stromatolite dolostone.

The middle part characterized by dark grey stromatolite dolostone, light grey dolostone with many siliceous bands, grey-purple and purple-red stromatolite dolostone, as well as purple-red copper-bearing stromatolite dolostone.

The upper part characterized by laminated variegated colour dolostone and light grey to flesh pink dolostone interbeds, black grey siliceous dolomitic siltstone, and black carbonaceous dolomitic siltstone.

The sedimentary structures in this formation are dominated by banded horizontal bedding, along with cross bedding, ripple marks, mud cracks, oolites, gravity gliding, fractures, injection dykes, syngenetic clasts and breccias.

图 2.7.1
(摄影:钱迈平)

Fig. 2.7.1
(Photograph by Qian Maiping)

官门山公路沿线：

石槽河组下部紫红色中薄层状含铜纹层白云岩，局部可见蓝绿色孔雀石粉末顺层间裂隙分布。孔雀石是碱式碳酸铜$Cu_2(OH)_2CO_3$，由铜与空气中的氧气、二氧化碳和水等物质反应产生，又称铜锈。因此处于碳酸盐台地潮坪下部，水体较平静，沉积物细腻，纹层发育；沉积岩石显示的紫红色，反映了当时炎热潮湿气候下的浅海氧化环境。

Along Guanmenshan highway:

Purple red medium- to thin-bedded copper-bearing laminated dolostones with visible locally blue-green malachite powder in interlayer fissures in the lower part of the Shicaohe Formation. Malachite is a copper carbonate hydroxide mineral, with the formula $Cu_2(OH)_2CO_3$, and results from the weathering of copper ores, when copper reacts with oxygen, carbon dioxide and water in air. The laminated dolostones composed of silt and clay particles slowly deposited through suspension in a calm subtidal zone of the carbonate platform, and its purple red color indicates deposition in an oxidizing shallow marine environment under alternating hot and humid climate.

图 2.7.2
(摄影:钱迈平)

Fig. 2.7.2
(Photograph by Qian Maiping)

石槽河组下部中薄层状紫红色泥质纹层白云岩与粉黄色白云质泥岩不等厚互层,重力滑动、拉张断裂及阶梯状注射脉构造发育。显示在干旱-湿润季节交替的萨布哈盐坪环境,斜坡上交替沉积的碳酸盐层与泥质层在受到扰动(风暴或潮汐海浪)或振动(火山喷发或地震)时,因重力作用而向下坡滑动,其中碳酸盐层发生断裂,裂隙被泥质注射脉充填。其过程是:致密而不透水的碳酸盐层断裂时,下伏松散而尚未固结并饱含水分的泥质层在上覆碳地层重压下,呈流体向上注射充填上覆地层的裂隙,形成注射脉,并随着地层的重力滑动而错开成阶梯状。

Alternating beds of the medium- to thin-bedded purple red muddy laminated dolostones and light pink yellow dolomitic marlstones, with gravity gliding, fracturing and injection dyke structures in the lower part of the Shicaohe Formation. Marl dykes associated with syndiagenetic normal faults, cutting through the alternating beds and subsequently separating along the gliding direction. It suggested that the alternating carbonate and marl beds on a slope occur gravity gliding and fractures in a sabkha environment alternated between dry and wet seasons. When a disturbance or shaking (such as a storm, volcanic eruption or earthquake) occurred, the dense, impermeable carbonate beds are fractured, meanwhile the underlying unconsolidated, loosely packed and water-saturated marls may rise through the fissures and flow over the carbonate beds.

图 2.7.3 （摄影：钱迈平）　　　　　　　　　　Fig. 2.7.3 （Photograph by Qian Maiping）

石槽河组下部浅灰色中厚层状含铜叠层石生物礁白云岩。以朱鲁莎叠层石（*Jurusania*）为主构成生物礁，叠层石基本层间裂隙间有绿色孔雀石粉末，显示当时此处处于碳酸盐台地潮坪，光照充足，叠层石发育良好。

Light grey medium- to thick-bedded copper-bearing stromatolite dolostones in the lower part of the Shicaohe Formation. The stromatolite bioherms composed mainly of *Jurusania* with blue green malachite powder in cracked laminae, formed on a sunny tidal flat of the carbonate plateform.

图 2.7.4
(摄影:钱迈平)

Fig. 2.7.4 (Photograph by Qian Maiping)

　　石槽河组中部深灰色块状叠层石生物礁白云岩,夹燧石透镜体及条带。以层叠层石(*Stratifers*)为主构成生物礁,显示当时处于碳酸盐台地潮坪,光照充足,叠层石非常发育。

　　在这面挂满蜂箱的岩壁上,可见层层延续的古老叠层石形成的巨大生物礁,以及来自当时火山活动的硅质热液沉积燧石透镜体及条带。

Dark grey massive stromatolite dolostone with black chert lens and bands in the middle part of the Shicaohe Formation.

The stromatolite biostromes composed mainly of *Stratifers* formed on a sunny tidal flat of the carbonate plateform. This steep cliff 1,200 meters above sea level on a mountain is a giant ancient biostrome made by layer upon layer of stromatolites, with chert deposits from ancient volcanic hydrothermal fluid. Present-day beekeepers pinned over 700 beehives on this cliff face to provide a habitat for the native bee population.

图 2.7.5
(摄影：钱迈平)

Fig. 2.7.5
(Photograph by Qian Maiping)

灰黑色与灰白色交替叠加增长的纹层是远古潮坪微生物席周期性生长留下的遗迹。

The alternating grey black and grey white laminae were the remains of ancient microbial mats on tidal flat grown atop older layers cyclically.

图 2.7.6 （摄影：钱迈平）   Fig. 2.7.6 （Photograph by Qian Maiping）

石槽河组中部浅灰色薄层－块状硅质条带白云岩。显示当时此处水深有所加大，水体较平静。火山作用的硅质热液沉积发育。

Light grey thin-bedded to massive dolostones with siliceous bands in the middle part of the Shicaohe Formation. It indicated the depth of water increased and transformed into a relatively calm environment with siliceous hydrothermal deposits from volcanisms.

图 2.7.9 （摄影：钱迈平）   Fig. 2.7.9 （Photograph by Qian Maiping）

  波层叠层石（*Stratifera undata*），基本层横向延展范围大，以平缓波状起伏为主，时而出现丘状突起。体现当时水很浅，光照充足，微生物席可绵延伸展，局部的微生物群落还表现出较强的趋光行为。

Stratifera undata, laminae extended laterally very well, and are generally moderately convex, sometimes occurred dome-shaped protuberances. It indicates the stretching ancient microbial mats developed in well-illuminated shallow water, with partial communities expressed a strong positive phototaxis.

图2.7.10 (摄影:钱迈平)　　　　　　　　　Fig. 2.7.10　(Photograph by Qian Maiping)

贝加尔贝加尔叠层石(*Baicalia baicalica*),基部收缩,散开分枝。显示当时水很浅,光照充足,各微生物群落可充分展开,享受阳光。

Baicalia baicalica, the columns are divergent branching tuberous with restrictions at the base, and showed the ancient microbial mats stretched very well and enjoyed sunshine in shallow water.

图 2.7.11 （摄影：钱迈平）　　　　　　　　Fig. 2.7.11　（Photograph by Qian Maiping）

石槽河组中部紫红色中薄层状含铜叠层石生物礁白云岩。以圆柱朱鲁莎叠层石（*Jurusania cylindrica*）为主构成生物礁，局部见铜铁硫化物粉末沿基本层风化面分布，叠层石柱体竖直生长，平行排列，显示当时处于碳酸盐台地潮坪或潟湖，叠层石发育良好。

Purple red medium- to thin-bedded copper-bearing stromatolite dolostone in the middle part of the Shicaohe Formation. The bioherms were consisted mainly of *Jurusania cylindrical* with partial copper iron sulfide in weathered laminae, and the parallel columns suggested that they grew well in a carbonate tidal flat or lagoon.

石槽河组下—中部紫红色白云岩的出现，说明当时位于滨海环境，如碳酸盐台地、潮坪、潟湖及萨布哈盐坪，气候炎热，水体流畅，光照充足，氧气丰富，叠层石生物礁发育，类型包括圆柱朱鲁莎叠层石（*Jurusania cylindrica*）、奥姆泰尼奥姆泰尼叠层石（*Omachtenia omachtensis*）、贝加尔贝加尔叠层石（*Baicalia baicalica*）和波层叠层石（*Stratifera undata*）等。同时，受火山作用影响，热液沉积发育。含铜的纹层白云岩和叠层石生物礁白云岩的出现，显示当时海底微生物群落的生命活动，对热液携带的某些金属矿物有富集作用。随着海水的变深，石槽河组上部灰黑—黑色含硅质、碳质的白云质粉砂岩的出现，说明当时此处已转变成较闭塞或较深的海盆，水体停滞，光照不足，氧气缺乏，叠层石消失。

In the lower-mid parts of the Shicaohe Formation, purple-red dolostones indicated they located in a carbonate platform, tidal flat, lagoon and sabkha environments, where the climate was hot, and the water was shallow, fluent, abundant sunshine and oxygen-rich. The well-developed stromatolites included *Jurusania cylindrica*, *Omachtenia omachtensis*, *Baicalia baicalica* and *Stratifera undata* etc. The hydrothermal sediments from volcanism, and copper-bearing laminated, stromatolite dolostones indicated the benthic microbial communities activities contributed to the enrichment and precipitation of some metallic minerals from hydrothermal fluids. In the upper part of the Shicaohe Formation, black grey siliceous and black carbonaceous dolomitic siltstones showd a bad light anoxybiotic stagnant environment with water was getting deeper and deeper. The stromatolites disappeared there.

2.8 神农架叠层石大堡礁的终结

2.8 THE END OF STROMATOLITIC GREAT BARRIER REEFS IN SHENNONGJIA

图 2.8.1 （摄影：钱迈平）　　　　　　　　Fig. 2.8.1　（Photograph by Qian Maiping）

靠近凉风垭的公路沿线，出露了厚度数以千米计的不等厚互层的粉砂岩和砂岩。显示当时随着整个区域的海水加深，碳酸盐台地没有了，叠层石也随之消失，沉积物以大量的从陆地岩石风化剥蚀下来的各种碎屑为主。反映了当时邻近的陆地在强烈构造运动作用下，不断抬升并遭受强烈风化剥蚀；同时这里的海底不断下沉，接受着陆地风化剥蚀下来的各种碎屑沉积，形成了粉砂岩和砂岩交替出现的巨厚地层。这到底是怎么回事呢？原来这是古地理发生了重大改变。

The thousands of meters thick rock sequences exposed along the highway near Liangfengya are mainly interbeded with an unequal thickness of siltstones and sandstones. It indicated the carbonate platform where stromatolites formed on disappeared while sea water deepened, and the sediments are mainly debris eroded from land rocks with adjacent lands continuously uplifted and strongly weathered. What happened? This is because the paleogeography had dramatically changed.

图 2.8.2　[据Goodge et al. (2008)描绘：马雪]　　Fig. 2.8.2　[Redrew by Ma Xue based on Goodge JW et al. (2008)]

根据地质学和古地磁学研究,中元古代晚期,地球上曾存在过一个罗迪尼亚超级大陆(Rodinia supercontinent),是在约11亿年前由一个更古老的哥伦比亚超级大陆(Columbia supercontinent)解体后的大大小小古陆块碰撞拼合而成的。那时的神农架位于华南古陆沿海,与澳大利亚古陆和西伯利亚古陆相邻,都在罗迪尼亚超级大陆北端,其碳酸盐台地上的叠层石组合很相似。常见的叠层石包括:大型的锥叠层石(*Conophyton*)、圆柱叠层石(*Colonnella*)、中小型的包心菜叠层石(*Cryptozoon*)、库什叠层石(*Kussiella*)、印卓尔叠层石(*Inzeria*)、朱鲁莎叠层石(*Jurusania*)、贝加尔叠层石(*Baicalia*)、通古斯叠层石(*Tungussia*)以及层叠层石(*Stratifera*)等。

约11亿年前,正值又一轮全球性的大规模构造运动时期,造山运动此起彼伏,例如北美的格林威尔造山运动(Grenville Orogeny)、波罗的海地区的达尔斯兰造山运动(Dalslandian Orogeny),以及北欧的瑞典-挪威造山运动(Sveconorwegian Orogeny)。不过,这些大规模造山运动对神农架地区影响不大。

然而,约10亿年前,华南的四堡造山运动(Sipu Orogeny)波及到神农架及其周围地区,地层受挤压作用发生褶皱。有的地区陆地抬升,因那时还没有陆地植物,光秃秃的陆地风化剥蚀非常强烈,时刻都在产生着大量

According to geological and paleomagnetic studies, there was a supercontinent, Rodinia, formed about 1,100 million years ago by the accretion and collision of fragments produced by breakup of an older supercontinent, Columbia. At that time, the Shennongjia in the South China craton adjoining Australian and Siberian cratons located northern Rodinia supercontinent. Similar stromatolite assemblages were well developed in coastal carbonate plateforms of these cratons. The common stromatolites included *Conophyton*, *Colonnella*, *Cryptozoon*, *Kussiella*, *Inzeria*, *Jurusania*, *Baicalia*, *Tungussia* and *Stratifera* ect.

Then a series of large-scale orogenies rising here and falling there, such as Grenville Orogeny in North America, Dalslandian Orogeny in Baltica and Sveconorwegian Orogeny in Northern Europa, had little influence to Shennongjia.

About 1,000 million years ago, however, the Sipu Orogeny in South China affected strongly the Shennongjia and surrounding areas. The formations folded and faulted by squeezing caused some regions uplifting

的岩石碎屑;同时,有的地区海底下沉,不断接受这些来自陆地的岩石碎屑,形成了由砂岩、粉砂岩和泥岩组成的巨厚地层。

约8亿年前,随着江南造山带的形成,神农架地区被整体抬升出水,完全成为陆地。

到7.5亿—6亿年前的新元古代晚期时,罗迪尼亚超级大陆已分裂成许多大大小小的古陆块。如图2.8.2所示,其中绿色是11亿年前的造山带,红色是造山运动结束后岩石圈拉张形成的A型花岗岩出露地点,粉色是古陆块,蓝色是古海洋。

into bare lands, without any land plants at that time, so these lands were strongly weathered and eroded, and a great many rock detritus were created; some regions subsiding and increasing depth of water, and depositing massive thick layers of rock detritus from lands.

About 800 million years ago, The Shennongjia was thoroughly uplifted into a land, with the Jiangnan orogenic belt formed.

In the late Neoproterozoic, about 750－600 million years ago, Rodinia Supercontinent broke up and split into several pieces as the above figure shows. The proposed reconstruction of Rodinia 750 million years ago, with 1,100 million years old orogenic belts highlighted in green. Red dots indicate 1,300－1,500 million years old A-type granites. Light pink areas and light blue areas represent separately cratons and seas.

图 2.8.3 （图片来源：pixnio.com） Fig. 2.8.3 （Photo gredit: pixnio.com）

约 7.2 亿年前，地球历史进入成冰纪（Cryogenian），发生多次全球性大冰期。年均气温低到-50 ℃，赤道海面冰层可超过 2 km！这就是著名的"雪球地球（The Snowball Earth）"时期。由于巨量的水在陆地上形成大规模冰川，海平面急剧下降，陆地扩大。已处于内陆的神农架地区只遭受剥蚀，没有沉积。

About 720 million years ago, the Cryogenian Period began, the name of the geologic period refers to the very cold global climate, and may have occurred several ice ages and produced a Snowball Earth in which perhaps the average temperature on the surface of Earth was only about -50 degree Celsius, and the sea ice on the equator was more than 2 kilometers thick! Ice ages tied up huge volumes of water in continental ice sheets, and the sea-level dropped dramatically, and meanwhile the continents extended drastically. The Shennongjia was located inland at that time, and suffered severe eroding without any sedimenting.

图 2.8.4 （摄影：钱迈平）　　　　　　　　　　Fig. 2.8.4　（Photograph by Qian Maiping）

  约 6.8 亿年前，"雪球地球"进入后期，由于冰雪的覆盖阻止了二氧化碳对岩石的风化，而风化又是消耗这种从火山源源不断释放出的气体的关键，结果二氧化碳在大气中积累到极其高的浓度，产生极其强大的温室效应导致出现高温酷热气候，冰层也随之融化而产生巨量的水，又导致了海平面大幅度上升。陆地冰川融化的水，通过迅速发达起来的河流水系，携带着剥蚀的大量岩石碎屑涌向湖泊或海洋，在一些地区形成洪积砂砾岩及河流相粗砂岩。另一方面，正在融化着的陆地冰川从高处向低处移动，携带着沿途刨蚀的大量岩石碎屑进入湖泊或海洋，漂浮在水面并逐渐消融，将携带的岩石碎屑也沉积下来，形成泥砂质冰碛砾岩。此时，在华南依次沉积形成了：莲沱组河流相粗砂岩、古城组泥砂质冰碛砾岩、大塘坡组锰矿层或新余式铁矿层，以及南沱组泥砂质冰碛砾岩等，显示至少曾有两次融冰期。

  随着全球冰层的融化，神农架地区再次被淹没在水下，并接受了融冰带来的大量沉积。在神农架凉风垭公路边可见，**南沱组**底部灰紫—浅灰色中薄层状泥砂质冰碛砾岩，与下覆地层**莲沱组**顶部浅灰绿—暗紫色凝灰质片理化粗砂岩假整合接触。

About 680 million years ago, the late stages of the Snowball Earth, the covering of ice and snow had stopped rocks being weathered by carbon dioxide for millions of years: weathering is the key process that uses up carbon dioxide which is continuously released into the atmosphere from volcanoes. So, the carbon dioxide progressively rose to unusually high levels, and caused the extreme strong greenhouse effect and extreme hot climate. As ice and snow melted and produced huge amounts of water, leading to a significant rise in sea level. The running water carrying the particles washed off of eroding rocks flowed to the sea and lakes through rapidly developing river systems from melting glaciers and ice caps on the land. When the energy of the transporting current was not strong enough to carry these particles, the particles dropped out in the process of sedimentation, and formed finally fluvial conglomerates and sandstones. Thus the sequences formed in South China in an ascending order: the coarse-grained fluvial sandstone of the Liantuo Formation, the glacial mud-sand matrix conglomerate of the Gucheng Formation, the manganese deposits of the Datangpo Formation or Xinyu-type iron deposits, as well as the glacial mud-sand matrix conglomerate of the Nantuo Formation etc. These indicted at least two times of melting glaciers.

As glaciers around the world melted, the Shennongjia was submerged again and accumulated large amount of sediments transported and deposited by ice and snow melting water flow. Along the Liangfengya highway, you can see the contact of grey purple-light gray mid-thin bedded glacial mud and sand matrix conglomerate in the bottom of the **Nantuo Formation** and the underlying light grey green-dark purple foliated tuffaceous coarse sandstone in the top of the **Liantuo Formation** is a disconformity.

图2.8.5 （摄影：钱迈平）　　　　　Fig. 2.8.5 （Photograph by Qian Maiping）

南沱组泥砂质冰碛砾岩和莲沱组凝灰质粗砂岩，分别呈现的灰紫色和暗紫色反映了"雪球地球"末期极度酷热的气候。它们超覆在神农架群地层之上，与该群不同的岩石地层组呈角度不整合接触。显示了约10亿年前的四堡造山运动的影响，将神农架地区的岩石地层挤压褶皱，而约8亿年前的江南造山带的形成，将神农架地区完全抬升成为陆地后，中断沉积超过1.6亿年，直到"雪球地球"行将结束时，才再次接受沉积。

The glacial mud and sand matrix conglomerate of the Nantuo Formation and the tuffaceous coarse sandstone of the Liantuo Formation appeared separately grey purple and dark purple colors, reflected an extreme hot climate in terminal Snow Earth. The Liantuo Formation is overlying an angular unconformity on the formations of Mesoproterozoic Shennongjia Group. It shows that Shennongjia Group was squeezed and folded during the Sipu Orogeny about 1,000 million years ago, subsequently the Jiangnan orogenic belt formed about 800 million years ago, the Shennongjia was thoroughly uplifted into a land and interrupted deposition for more than 160 million years until the Snow Earth melted.

神农架地质公园主要景区出露的中元古系神农架群地层,自下而上反映了沉积环境的演变:

乱石沟组、大窝坑组及矿石组地层,沉积于从盆地边缘到碳酸盐台地之间反复变化的环境,但总的趋势是水深变浅。在盆地环境,水流不畅而多黑色泥质碳酸盐沉积,受海底火山作用影响因而硅质及铁质碳酸盐沉积发育;在碳酸盐台地边缘斜坡环境,多见重力流、重力褶皱及重力滑动等构造;在碳酸盐台地边缘礁滩、潮坪及潟湖环境,叠层石生物礁大量形成。

台子组地层沉积于由碳酸盐台地向盆地边缘转变的环境。在台地潮坪、潟湖及边缘礁滩环境,叠层石生物礁群很发育;在斜坡环境,多重力流形成砾岩层;在盆地边缘环境,以黑泥质碳酸盐沉积为主。

野马河组及温水河组地层沉积于由盆地边缘向碳酸盐台地转变的环境,随着海水的变浅,叠层石生物礁也随之出现。

Mesoproterozoic Shennongjia Group sequences exposed along highways in main tourist spots of Shennongjia UNESCO Global Geopark reflects the evolution of sedimentary environments in an ascending order:

The Luanshigou, Dawokeng and Kuangshishan formations were deposited in a changing repeatedly environment from a basin margin to a carbonate platform, and the general trend was the water depth getting shallower. In a basin margin condition, black muddy carbonate deposits were common in a stagnant water, and siliceous or iron carbonate deposits were formed as a result of submarine volcanism. In a carbonate platform margin slope condition, sediment of gravity flows, gravity folds, gravity gliding and gravity spreading were often observed. In a carbonate platform margin-shoal or tidal flat or lagoon condition, stromatolite biostromes and bioherms were well developed.

The Taizi Formation was deposited during a transition period from a carbonate platform to a basin margin. Stromatolite biostromes and bioherms bloomed in a tidal flat or lagoon or carbonate platform margin-shoal condition. Sediment of gravity flows, gravity folds, gravity gliding and gravity spreading were frequently occurred in a marginal slope condition. Black muddy carbonate deposits precipitated mainly in a basin margin condition.

石槽河组地层沉积于碳酸盐台地潮坪向盆地转变期间，海水趋向变深。在潮坪环境，形成叠层石生物礁群；在盆地环境，没有叠层石，沉积黑泥及粉砂。

　　从神农架群地层的形成过程可见，在12亿多年前的中元古代中、晚期，处于炎热气候下的华南古陆沿海，在碳酸盐台地潮坪、潟湖及边缘礁滩发育大量叠层石生物礁，局部可形成延绵几十千米的由叠层石生物礁群组成的大堡礁，其中以大窝坑组、矿石山组及台子组下部的叠层石大堡礁最发育。此后海侵加强，海水呈变深趋势。最终在约10亿年前的新元古代初期，叠层石生物礁消失。

The Yemahe and Wenshuihe formations were deposited during a transition period from a basin margin to a carbonate platform. Stromatolite biostromes and bioherms were growing well as the water depth getting shallower.

The Shicaohe Formation were deposited during a transition period from a carbonate platform to a basin. Stromatolites disappeared as the water depth getting deeper.

In summary, the Shennongjia Group sequences were deposited in coastal areas of the South China craton in a hot climate during the mid-late Mesoproterozoic, more than 1,200 million years ago. The extensive stromatolite biostromes and bioherms occurred commonly in carbonate platform tidal flats, lagoons and margin-shoals, and some of the bioherm groups stretched several dozens or hundreds kilometers were almost comparable to modern Great Barrier Reef. In which stromatolite biostromes and bioherms were most widely developed in the Dawokeng Formation, the Kuangshishan Formation and the lower part of the Taizi Formation. With subsequent marine transgression, the water got deeper and deeper until the stromatolites cannot form in early Neoproterozoic, about 1,000 million years ago.

3 神农架中元古代叠层石

3 MESOPROTEROZOIC STROMATOLITES FROM SHENNONGJIA

中元古代时期,神农架位于华南古陆,与澳大利亚古陆及西伯利亚古陆相邻,同处于罗迪尼亚超级大陆(Rodinia supercontinent)北部。其沿海的叠层石组合,与澳大利亚古陆及西伯利亚古陆沿海的叠层石组合相似,显示了炎热气候下的碳酸盐台地潮坪浅水环境特征。

常见的叠层石形态类型包括:穹状、柱状、层柱状和层状等。主要有:神农架大圆顶叠层石(*Megadomia shennongjiaensis*)、加尔加诺锥叠层石(*Conophyton garganicum*)、树桩圆柱叠层石(*Colonnella cormosa*)、简单包心菜叠层石(*Cryptozoon haplum*)、喀什喀什叠层石(*Kussiella kussiensis*)、圆柱朱鲁莎叠层石(*Jurusania cylindrica*)、地窖印卓尔叠层石(*Inzeria intia*)、瘤通古斯叠层石(*Tungussia nodosa*)、贝加尔贝加尔叠层石(*Baicalia baicalica*)、育卡贝加尔叠层石(*Baicalia unca*)、奥姆泰尼奥姆泰尼叠层石(*Omachtenia omachtensis*)和波层叠层石(*Stratifera undata*)等。

During Mesoproterozoic Era, Shennongjia region located in South China Craton, adjoining Australia Craton and Siberia Craton, in northern Rodinia Supercontinent. The stromatolites in off the coasts of these continents were similar in assemblages. They formed in shallow water carbonate platform and tidal flat in a hot climate.

The common morphological types of the stromatolites in Shennongjia region include: domed, columnar, columnar-stratiform and stratiform stromatolites etc. Such as *Megadomia shennongjiaensis*, *Conophyton garganicum*, *Colonnella cormosa*, *Cryptozoon haplum*, *Kussiella kussiensis*, *Jurusania cylindrica*, *Inzeria intia*, *Tungussia nodosa*, *Baicalia baicalica*, *Baicalia unca*, *Omachtenia omachtensis* and *Stratifera undata* etc.

图 3.1.1 （摄影：钱迈平）　　　　　　　　Fig. 3.1.1　（Photograph by Qian Maiping）

神农架大圆顶叠层石（*Megadomia shennongjiaensis*）纵断面。
化石层位：中元古代矿石山组
化石地点：神农架林区神农顶－凉风垭公路边

Megadomia shennongjiaensis, longitudinal section.

Horizon: Mesoproterozoic Kuangshishan Formation

Locality: Shennongding-Liangfengya highway, Shennongjia Forestry District, Central China

图 3.1.2 （摄影：钱迈平）　　　　　　　　Fig. 3.1.2 （Photograph by Qian Maiping）

神农架大圆顶叠层石（*Megadomia shennongjiaensis*）相邻叠层体交错重叠生长。

化石层位：中元古代矿石山组

化石地点：神农架林区神农顶－凉风垭公路边

Giant domed stromatolites *Megadomia shennongjiaensis* overlapped alternately.

Horizon: Mesoproterozoic Kuangshishan Formation

Locality: Shennongding-Liangfengya highway, Shennongjia Forestry District, Central China

图 3.1.3 （摄影：钱迈平）　　　　　　　　Fig. 3.1.3 （Photograph by Qian Maiping）

巨大的神农架大圆顶叠层石（*Megadomia shennongjiaensis*）高、宽达数米。

化石层位：中元古代矿石山组
化石地点：神农架林区神农顶—凉风垭公路边

The giant domal structures of *Megadomia shennongjiaensis* are seversl meters wide and high.

Horizon: Mesoproterozoic Kuangshishan Formation
Locality: Shennongding-Liangfengya highway, Shennongjia Forestry District, Central China

图 3.1.4　（摄影：钱迈平）　　　　　　　Fig. 3.1.4　（Photograph by Qian Maiping）

巨大的神农架大圆顶叠层石（*Megadomia shennongjiaensis*）基部密集并列生长的小叠层石。

化石层位： 中元古代矿石山组

化石地点： 神农架林区神农顶－凉风垭公路边

The base of a giant domed stromatolite *Megadomia shennongjiaensis* formed by a cluster of smaller column stromatolites.

Horizon: Mesoproterozoic Kuangshishan Formation

Locality: Shennongding-Liangfengya highway, Shennongjia Forestry District, Central China

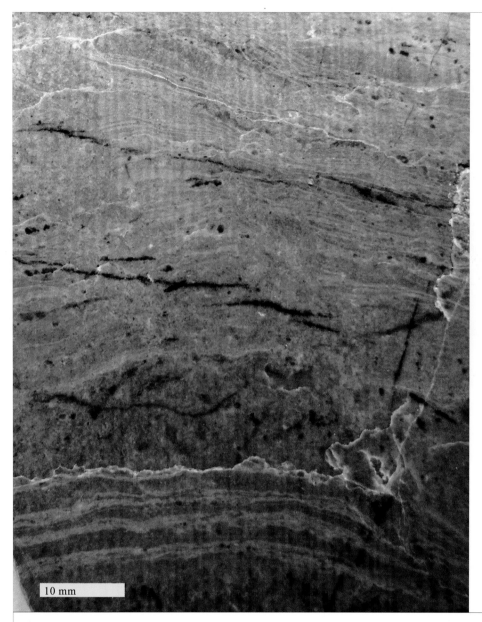

图 3.1.5 （摄影：钱迈平）　　　　　　　　　Fig. 3.1.5 （Photograph by Qian Maiping）

神农架大圆顶叠层石（*Megadomia shennongjiaensis*）的基本层纵切面，显示古微生物席的有机质降解后形成的暗层与碳酸盐沉积结晶形成的亮层，交替叠加组成不规则起伏并大致平行的叠层构造。

A polished longitudinal section of *Megadomia shennongjiaensis* shows the alternating dark ancient microbiota mat remain laminae and light carbonate crystal laminae that are irregularly undulant and roughly parallel.

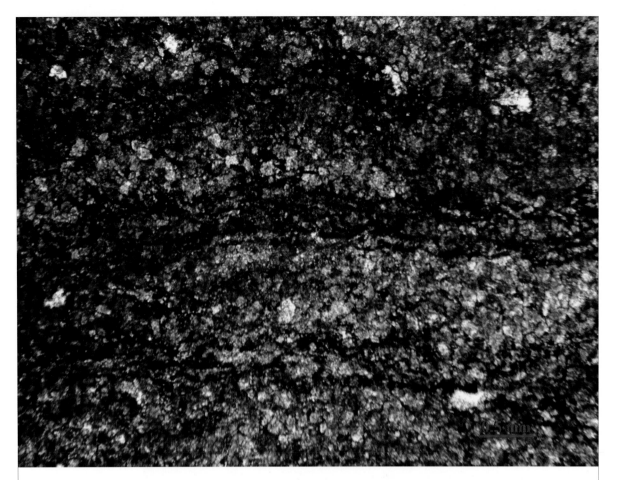

图 3.1.6 （摄影：马雪） Fig. 3.1.6 （Photograph by Ma Xue）

神农架大圆顶叠层石（*Megadomia shennongjiaensis*）的基本层纵切面微结构，显示丝状交织的微生物席化石结构。

The microstructure of *Megadomia shennongjiaensis*: Alternation of sparry laminae and tangle wire-like fossilized microbial mat laminae, with indistinct boundaries and varying continuity.

3.2 简单包心菜叠层石
(*Cryptozoon haplum* Liang, 1979)

分类：

球叠层石纲 Sphaerati Cao et Yuan, 2006
  包心菜叠层石科 Cryptozoonaceae Cao et Yuan, 2006
    包心菜叠层石属 Cryptozoon Hall, 1883
      简单包心菜叠层石 *Cryptozoon haplum* Liang, 1979

同义名：

1979 *Cryptozoon haplum* Liang,国家地质总局天津地质矿产研究所、中国科学院南京地质古生物研究所、内蒙古自治区地质矿产局,蓟县震旦亚界叠层石的研究,地质出版社,83页;图版43:图1—5。

描述： 小型扁椭球状叠层体,不分枝,直径4—6 cm,高2—3 cm。基

3.2 *Cryptozoon haplum* Liang, 1979

Systematics:

Class: Sphaerati Cao et Yuan, 2006
Family: Cryptozoonaceae Cao et Yuan, 2006
Genus: *Cryptozoon* Hall, 1883
Species: *Cryptozoon haplum* Liang, 1979

Content:

1979 *Cryptozoon haplum* Liang, Tianjin Institute of Geology and Mineral Resources, Nanjing Institute of Geology and Palaeontology, The Inner Mongolia Autonomous Region Bureau of Geology and Mineral Resources, Jixian County Sinian Suberathem Stromatolite Research. Geological Publishing House, Beijing: p. 83, plate 43: figs. 1—5.

Description: Non-branching oblate ellipsoidal stromatolites with a base consists of a cluster of small non-branching oblate ellipsoidal

部多个小扁椭球形叠层体,不分枝,被外面的共同层包裹,每个小叠层体直径1 cm左右,基本层穹形,一层层紧密包裹。带状或海绵状微结构。

分布及层位:神农架林区神农顶－凉风垭,中元古代大窝坑组;天津蓟县下营团山子,古元古代团山子组等。

structures. Each small oblate ellipsoidal structure is about 1 cm in diameter and these structures are wrapped together by common layers, passing up into a large structure up to 4－6 cm in diameter and 2－3 cm high. The laminar microstructures are ribbon-like or sponge-like.

Locality and horizon: Mesoproterozoic Dawokeng Formation, Ectasian, Shennongding-Liangfengya highway, Shennongjia Forestry District, Central China; Paleoproterozoic Tuanshanzi Formation, Tuanshanzi, Xiaying Town, Jixian County, Tianjin City, North China, etc.

图 3.2.1 （摄影：钱迈平） Fig. 3.2.1 (Photograph by Qian Maiping)

简单包心菜叠层石（*Cryptozoon haplum*）纵断面。
化石层位：中元古代大窝坑组
化石地点：神农架林区神农顶－凉风垭公路边

Megadomia shennongjiaensis, longitudinal section
Horizon: Mesoproterozoic Dawokeng Formation
Locality: Shennongding-Liangfengya highway, Shennongjia Forestry District, Central China

3.3 加尔加诺锥叠层石
(*Conophyton garganicum* Koroljuk, 1963)

分类:
柱叠层石纲 Columellati Cao et Yuan, 2006
  不分枝柱叠层石目 Unramificolumllati Cao et Yuan, 2006
    锥叠层石科 Conophytonaceae Raaben, 1969
      锥叠层石属 *Conophyton* Maslov, 1937
        加尔加诺锥叠层石 *Conophyton garganicum* Koroljuk, 1963

同义名:

1963 *Conophyton garganicus* Koroljuk, Koroljuk I K, In: Keller B M, ed, Stratigrafiya USSR, Moscow, pp. 479—498; plate Ⅴ: fig. 3.

1965 *Conophyton garganicus*, Komar V A, Raaben M E, Semikhatov M A, Academy of Sciences of the USSR, Geol. Inst., Transactions, 131, pp. 42—46; plate Ⅷ: figs. 1—3.

1969 *Conophyton* cf. *garganicus*, Hofmann H J, Geol. Surv. pap. Can., pp. 68—69.

1969 *Conophyton* cf. *garganicus*, Glaessner M F, *et al.*, Science, 164 (3883), pp. 1056—1058; text-figs. 2, 3.

1969 *Conophyton garganicus*, Cloud P E, Semikhatov M A, Amer. Jour. Sci., 267 (9), p. 1038; plate 1: figs. 3, 4; plate 2: fig. 1.

1972 *Conophyton garganicum* ? *garganicum*, Walter M R, Special papers in palaeontology (11), published by the Palaeontological Association, pp. 110—111; plate 5: fig. 4; plate 11: fig. 3; plate 12: fig. 1.

1974 *Conophyton garganicus*,曹瑞骥,梁玉左,中国科学院南京地质古生物研究所集刊(5),图版Ⅴ:图3,5.

1979 *Conophyton garganicus*,国家地质总局天津地质矿产研究所、中国科学院南京地质古生物研究所、内蒙古自治区地质局,蓟县震旦亚界叠层石研究,地质出版社,74页;图版30:图1—3.

1980 *Conophyton garganicus*,梁玉左,地层古生物论文集:第8辑,地质出版社,28—29页;图版11:图4,5.

1982 *Conophyton garganicum*,张录易等,西北地区古生物:陕甘宁分册(一),地质出版社,352—353.

1993 *Conophyton garganicum*,缪长泉,新疆昆仑山和阿尔金山前寒武系及叠层石,新疆科技出版社(K),86—87页,图版1:图4,5;图版2:

3.3 *Conophyton garganicum* Koroljuk, 1963

Systematics:
Class: Columellati Cao et Yuan, 2006
Order: Unramificolumllati Cao et Yuan, 2006
Family: Conophytonaceae Raaben, 1969
Genus: *Conophyton* Maslov, 1937
Species: *Conophyton garganicum* Koroljuk, 1963

Content:

1963 *Conophyton garganicus* Koroljuk, Koroljuk I K, In: Keller B M, ed, Stratigrafiya USSR, Moscow, pp. 479—498; plate Ⅴ: fig. 3.

1965 *Conophyton garganicus*, Komar V A, Raaben M E, Semikhatov M A, Academy of Sciences of the USSR, Geol. Inst., Transactions, 131, pp. 42—46; plate Ⅷ: figs. 1—3.

1969 *Conophyton* cf. *garganicus*, Hofmann H J, Geol. Surv. pap. Can., pp. 68—69.

1969 *Conophyton* cf. *garganicus*, Glaessner M F, *et al.*, Science, 164 (3883), pp. 1056—1058; text-figs. 2, 3.

1969 *Conophyton garganicus*, Cloud P E, Semikhatov M A, Amer. Jour. Sci., 267 (9), p. 1038; plate 1: figs. 3, 4; plate 2: fig. 1.

1972 *Conophyton garganicum* ? *garganicum*, Walter M R, Special papers in palaeontology (11), published by the Palaeontological Association, pp. 110—111; plate 5: fig. 4; plate 11: fig. 3; plate 12: fig. 1.

1974 *Conophyton garganicus*, Cao R J and Liang Y Z, Memoirs of Nanjing Institute of Geology and Palaeontology, No. 5, plate Ⅴ: figs. 3, 5.

1979 *Conophyton garganicus*, Tianjin Institute of Geology and Mineral Resources, Nanjing Institute of Geology and Palaeontology, The Inner Mongolia Autonomous Region Bureau of Geology and Mineral Resources, Jixian County Sinian Suberathem Stromatolite Research. Geological Publishing House, Beijing: p. 74, plate 30: figs. 1—3.

1980 *Conophyton garganicus*, Liang Y Z, Professional Papers of Stratigraphy and Palaeontology, No. 8, Geological Publishing House, Beijing: pp. 28—29, plate 11: figs. 4, 5.

1982 *Conophyton garganicum*, Zhang L Y, et al., Paleontological Atlas of Northwest China, Shaanxi, Gansu and Ningxia Volume, Part Ⅰ. Geological Publishing House, Beijing: pp. 352—353.

图 1.

描述：中—大型锥柱状叠层体，不分枝。相互紧密平行排列，间距不超过 5 cm，垂直于岩层分布，构成厚达 10 余米的生物礁层。锥柱体横断面呈圆、椭圆、次圆及次多角形。锥柱体直径大多约 60 cm，高达 200 cm；基本层陡锥形，锥顶角 50°—70°；基本层顶部出现弯曲形成的轴带，轴带宽 2.4—9.7 mm，常交替收缩和膨胀；基本层在锥柱体边缘变薄。锥柱体表面不平整，檐十分发育；相邻锥柱体或彼此分开，或部分连层相连。基本层呈亮带—暗带交替，亮带厚度较稳定，大多厚 0.05—0.10 mm，少数超过 0.15 mm；暗带断续延伸，断片长度 1—20 mm 不等，厚度也不稳定，大多 0.025—0.100 mm，少数达 0.150 mm。断续呈丝及团块状微结构。

分布及层位：神农架林区神农顶—凉风垭，中元古代台子组；天津蓟县下营大红峪沟，中元古代高于庄组；内蒙古冈德尔山—桌子山地区，中元古代王全口组；新疆北天山西段科古尔琴山，中元古代四台组；俄罗斯乌拉尔—西伯利亚地区，中元古代尤尔马提群（Yurmatia Group）—新元古代早期卡拉陶群（Karatau Group）；澳大利亚麦克阿瑟盆地，中元古代麦克阿瑟群（McArthur Group）；班杰冒盆地，中元古代班杰冒群（Bangemall Group）；印度阿尔莫拉地区（District Almora），中元古代冈果利哈特白云岩（Gangolihat Dolomites）等。

1993 *Conophyton garganicum*. Miao C Q, Precambrian System and Stromatolites in the Kunlun Mountains and Altun Mountains, Xinjiang, China. Xinjiang Science and Technology Publishing House, Urumqi: pp. 86—87, plate 1: figs. 4,5; plate 2: fig. 1.

Description: Big or middle non-branching columnar stromatolites with conical laminae, contorted in their crestal parts. Columns in vertically parallel arrangement, no more than 5 cm apart, formed bioherm or thick biostrome, perhaps more than 10 m thick. Transverse sections of columns round to oval or polygon, usually about 60 cm in diameter and up to 200 cm in high. In longitudinal axial sections laminae steeply conical, apical angle generally acute 50—70°, and laminae usually straight and parallel in longitudinal section, but in places bent downwards near the column margins. All laminae more or less contorted in crestal zone. Dark laminae arched up and contorted, often leaving irregular voids filled with sparry dolomite, within the thickened light laminae. The crestal line, joining apices of successive conical laminae, is very wave, with frequent sharp displacements of crests. The diameter of the crestal zone is between 2.4 and 9.7 cm, with expanding and compressing alternately in diameter. The margin structure is very irregular, with numerous large bumps, overhanging peaks and short cornices. Bridges vary in thickness from one or two to several tens of laminae. Lamination is very distinctly banded and striated, consisting of straight, parallel, smooth, very thin laminae. The primary laminae occur light and dark layers altemately. The light layer is continuous and relatively pure transparent, mostly 0.05—0.10 mm, rarely more than 0.15 mm thick, generally of very constant thickness from the edge of the crestal zone to the column margin, never lensing out; while the dark layer is continuous or discontinuous or formed by chains of elongated lenses (1—2 mm long), mostly 0.025—0.100 mm, rarely 0.150 mm thick, of uneven thickness, with discontinuous wire-like and lumpy microstructures.

Locality and horizon: Mesoproterozoic Taizi Formation, Ectasian, Shennongding-Liangfengya highway, Shennongjia Forestry District, Central China; Mesoproterozoic Gaoyuzhuang Formation, Dahongyu, Xiaying Town, Jixian County, Tianjin City, North China; Mesoproterozoic Wangquankou Formation, Gangde'er Hill in Zhuozishan Mountain area, Inner Mongolia; Mesoproterozoic Sitai Formation, Keguerqin Mountain region in west segment of northern Tianshan Mountains, Xinjiang; Mesoproterozoic Yurmatia Group-Neoproterozoic Karatau Group, Urals-Siberia, Russia; Mesoproterozoic McArthur Group, McArthur Basin, Australia; and Mesoproterozoic Gangolihat Dolomites, District Almora, India etc.

图 3.3.1
(摄影:钱迈平)

Fig. 3.3.1
(Photograph by Qian Maiping)

加尔加诺锥叠层石(*Conophyton garganicum*)纵断面显示其锥状基本层在顶部出现弯曲。

化石层位: 中元古代台子组

化石地点: 神农架林区神农顶－凉风垭公路边

Conophyton garganicum, longitudinal section, shows its conical laminae contorted in their crestal parts

Horizon: Mesoproterozoic Taizi Formation

Locality: Shennongding-Liangfengya highway, Shennongjia Forestry District, Central China

图 3.3.2 （摄影：钱迈平） Fig. 3.3.2 （Photograph by Qian Maiping）

加尔加诺锥叠层石（*Conophyton garganicum*）基本层纵切面。

A polished longitudinal section of *Conophyton garganicum*.

图 3.3.3 （摄影:马雪） Fig. 3.3.3 （Photograph by Ma Xue）

加尔加诺锥叠层石（*Conophyton garganicum*）的基本层纵切面微结构，显示连续或断续呈丝及团块状的微生物席化石结构。

The microstructure of *Conophyton garganicum* shows thin laminae, either continuous or discontinuous, or formed by chains of elongated lenses and lumpiness, aligned in definite laminae.

3.4 树桩圆柱叠层石
(*Colonnella cormosa* Komar, 1964)

分类:

柱叠层石纲 Columellati Cao et Yuan, 2006
  不分枝柱叠层石目 Unramificolumllati Cao et Yuan, 2006
    锥叠层石科 Conophytonaceae Raaben, 1969
      圆柱叠层石属 *Colonnella* Komar, 1964
        树桩叠层石 *Colonnella cormosa* Komar, 1964

同义名:

1964 *Colonnella cormosa*, Komar V A, Academy of Sciences of the USSR, Geological Institute, Transactions, 154, pp. 69, 70; plate Ⅲ: figs. 1—3.

1979 *Colonnella* sp., 曹瑞骥,赵文杰,中国科学院铁矿地质学术会议论文选集

3.4 *Colonnella cormosa* Komar, 1964

Systematics:

Class: Columellati Cao et Yuan, 2006
Order: Unramificolumllati Cao et Yuan, 2006
Family: Conophytonaceae Raaben, 1969
Genus: *Colonnella* Komar, 1964
Species: *Colonnella cormosa* Komar, 1964

Content:

1964 *Colonnella cormosa*, Komar V A, Academy of Sciences of the USSR, Geological Institute, Transactions, 154, pp. 69, 70; plate Ⅲ: figs. 1—3.

1979 *Colonnella* sp., Cao R J and Zhao W J, Proceedings of the Symposium on Iron Ore Geology (1977), Chinese Academy of Sciences: Stratum and Paleontology. Science Press, Beijing: p. 79; plate Ⅳ: fig. 1; plate Ⅹ: fig. 5.

1982 *Colonnella* cf. *cormosa*, Zhang L Y et al., Paleontological Atlas of Northwest China, Shaanxi, Gansu and Ningxia Volume, Part Ⅰ Geological

（1977）；地层和古生物，科学出版社，79页，图版Ⅳ，图1；图版Ⅹ：图5。

1982 *Colonnella* cf. *cormosa*，张录易等，西北地区古生物图册，陕甘宁分册（一），地质出版社，358页；图版93：图4。

1991 *Colonnella cormosa*，李铨，冷坚，神农架上前寒武系，天津科学技术出版社，图版24：图4（未描述）。

描述：大—中型圆柱状叠层体，不分枝，相互紧密平行排列，间距不超过5 cm，垂直于岩层分布，构成厚达10余m的生物礁层。柱体横断面呈圆—次圆形，直径大多40－60 cm，高达200 cm，侧表面较光滑，未见檐或瘤；基本层平缓穹形，细密，厚0.2－0.5 mm，在柱体边缘下弯，但不构成明显的壁。呈不连续的丝状、斑点及团块状微结构。

分布及层位：神农架林区神农顶-凉风垭，中元古代矿石山组；辽宁大石桥南楼圣水寺，古元古代辽河群大石桥组；俄罗斯西伯利亚阿纳巴尔（Anabar），中元古代尤斯马斯塔克组（Yusmastak Formation）；印度阿尔莫拉地区（District Almora），中元古代冈果利哈特白云岩（Gangolihat Dolomites）等。

Publishing House, Beijing: p. 358; plate 93: fig. 4.

1991 *Colonnella cormosa*, Li Q and Leng J, The Upper Precambrian in the Shennongjia Region. Tianjin Science and Technology Press, Tianjin: plate 24: fig. 4 (undescribed).

Description: Big or middle non-branching columnar stromatolites with arched laminae. Columns in vertically parallel arrangement, no more than 5 cm apart, formed bioherm or thick biostrome, perhaps more than 10 m thick. Transverse sections of columns round or subround, mostly 40－60 cm in diameter and up to 200 cm in high. The lateral surface is fairly smooth, without a bump or overhanging peak. The primary laminae are fine and flat dome, 0.2－0.5 mm thick, in places bent downwards near the column margins, but does not form a distinct wall, with discontinuous wire-like, spots and chains of elongated lenses microstructures.

Locality and horizon: Mesoproterozoic Kuangshishan Formation, Ectasian, Shennongding-Liangfengya highway, Shennongjia Forestry District, Central China; Paleoproterozoic Dashiqiao Formation, Shenshuisi Village, Nanlou Subdistrict, Dashiqiao County-level City, Yingkou City, Liaoning Province; Mesoproterozoic Yusmastak Formation, Anabar region in northern Siberia, Russia; and Mesoproterozoic Gangolihat Dolomites, District Almora, India etc.

图 3.4.1 （摄影：钱迈平）　　　　　　　　Fig. 3.4.1 （Photograph by Qian Maiping）

树桩圆柱叠层石（*Colonnella cormosa*）横断面和纵断面。
化石层位：中元古代矿石山组
化石地点：神农架林区神农顶－凉风垭公路边

Colonnella cormosa, transverse and longitudinal sections.
Horizon: Mesoproterozoic Kuangshishan Formation
Locality: Shennongding-Liangfengya highway, Shennongjia Forestry District, Central China.

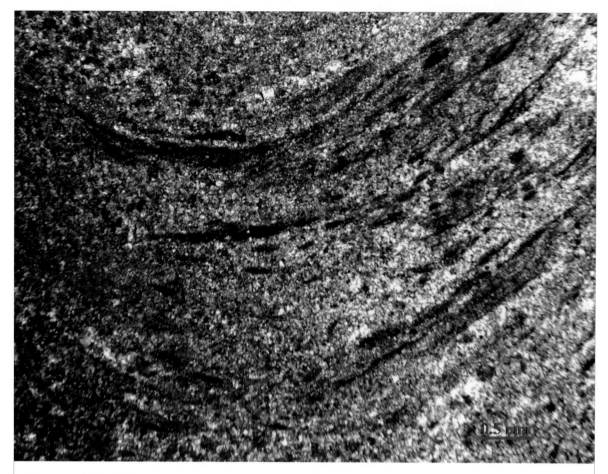

图 3.4.2 （摄影：马雪）　　　　　　　　Fig. 3.4.2　（Photograph by Ma Xue）

　　树桩圆柱叠层石（*Colonnella cormosa*）基本层横切面微结构，显示丝状、斑点及团块状的微生物席化石结构。

The microstructure of *Colonnella cormosa*, transverse sections, shows discontinuous wire-like, spots, lumpiness and chains of elongated lenses.

3.5 喀什喀什叠层石
(*Kussiella kussiensis*(Maslov)Krylov, 1963)

分类：

柱叠层石纲 Columellati Cao et Yuan, 2006

  分枝柱叠层石目 Ramificolumllati Cao et Yuan, 2006

    喀什叠层石科 Kussiellaaceae Raaben, 1969

      喀什叠层石属 *Kussiella* Komar, 1964

        喀什喀什叠层石 *Kussiella kussiensis*（Maslov）Krylov, 1963

同义名：

1939 *Collenia buriatica*, Maslov V P, Probl. Paleontol., 5: pp. 297－310; plate Ⅰ: fig. 1.

3.5 *Kussiella kussiensis*(Maslov)Krylov, 1963

Systematics:

Class: Columellati Cao et Yuan, 2006

Order: Ramificolumllati Cao et Yuan, 2006

Family: Kussiellaaceae Raaben, 1969

Genus: *Kussiella* Komar, 1964

Species: *Kussiella kussiensis* (Maslov) Krylov, 1963

Content:

1939 *Collenia buriatica*, Maslov V P, Probl. Paleontol., 5: pp. 297－310; plate Ⅰ: fig. 1.

1960 *Collenia kussiensis*, Krylov I N, Dokl. Akad., Nauk SSSR, 132 (4): pp. 895－896; plate Ⅰ: fig. a.

1960 *Collenia kussiensis*, Korolyuk I K, Tr. Geol. Inst. Akad., Nauk SSSR, 12: p. 35; plate Ⅰ: fig. 9; plate Ⅱ: fig. 1.

1963 *Kussiella kussiensis*, Krylov I N, Tr. Geol. Inst. Akad., Nauk

1960 *Collenia kussiensis*, Krylov I N, Dokl. Akad., Nauk SSSR, 132（4）: pp. 895−896; plate Ⅰ: fig. a.

1960 *Collenia kussiensis*, Korolyuk I K, Tr. Geol. Inst. Akad., Nauk SSSR, 12: p. 35; plate Ⅰ: fig. 9; plate Ⅱ: fig. 1.

1963 *Kussiella kussiensis*, Krylov I N, Tr. Geol. Inst. Akad., Nauk SSSR, 69: pp. 60−63; plate Ⅰ: fig. 5; text-figs. 16, 17.

1979 *Kussiella* cf. *kussiensis*,曹瑞骥,赵文杰,中国科学院铁矿地质学术会议论文选集（1977）;地层古生物,科学出版社,80页,图版Ⅶ:图4。

1982 *Kussiella* cf. *kussiensis*,张录易等,西北地区古生物:陕甘宁分册(一),地质出版社,p. 361,图版97:图1,2。

描述:次圆柱状叠层体,不增宽平行分枝,柱体垂直地层层理分布,紧密排列,间距通常不超过3 cm。柱体高度超过50 cm,横断面圆形至不规则形。母柱体大多宽12−14 cm,分出两个近等大的子柱体。子柱体大多宽6−7 cm,平行分布。柱体表面多明显的檐,基本层凸起平缓至揉曲状。模糊带状微结构。

分布及层位:神农架林区神农谷−神农顶,中元古代乱石沟组;辽宁大石桥南楼圣水寺,古元古代辽河群大石桥组;俄罗斯南乌拉尔,古元古代莎塔组(Satha Formation)等。

SSSR, 69: pp. 60−63; plate Ⅰ: fig. 5; text-figs. 16, 17.

1979 *Kussiella* cf. *kussiensis*, Cao R J and Zhao W J, Proceedings of the Symposium on Iron Ore Geology (1977), Chinese Academy of Sciences: Stratum and Paleontology. Science Press, Beijing: p. 80; plate Ⅶ: fig. 4.

1982 *Kussiella* cf. *kussiensis*, Zhang L Y, et al., Paleontological Atlas of Northwest China, Shaanxi, Gansu and Ningxia Volume, Part Ⅰ. Geological Publishing House, Beijing: p. 361, plate 97: figs. 1, 2.

Description: No widening parallel branching subcylindrical columns with rounded or irregular transverse sections. The columns in vertically parallel arrangement, no more than 3 cm apart, are mostly 12−14 cm in diameter and up to more than 50 cm high. Usually a column has two erect braches almost equaled in size which are mostly 6−7 cm in diameter. The lateral surface of the column is finely toothed, with numerous short cornices. Lamina profile varies from gentle to flexuous, with indistinct banding microstructures.

Locality and horizon: Mesoproterozoic Luanshigou Formation, Shennongding-Liangfengya highway, Shennongjia Forestry District, Central China; Paleoproterozoic Dashiqiao Formation, Shenshuisi Village, Nanlou Subdistrict, Dashiqiao County-level City, Yingkou City, Liaoning Province; Paleoproterozoic Satha Formation, southern Urals, Russia etc.

图 3.5.1 （摄影：钱迈平）　　　　Fig. 3.5.1　(Photograph by Qian Maiping)

喀什喀什叠层石（*Kussiella kussiensis*）。
化石层位：中元古代乱石沟组
化石地点：神农架林区神农谷－神农顶公路边

Kussiella kussiensis.
Horizon: Mesoproterozoic Luanshigou Formation
Locality: Shennonggu-Shennongding highway, Shennongjia Forestry District, Central China

3.6 圆柱朱鲁莎叠层石
(*Jurusania cylindrica* Krylov, 1963)

分类:

柱叠层石纲 Columellati Cao et Yuan, 2006
  分枝柱叠层石目 Ramificolumllati Cao et Yuan, 2006
    喀什叠层石科 Kussiellaaceae Raaben, 1969
      朱鲁莎叠层石属 *Jurusania* Krylov, 1963
        圆柱朱鲁莎叠层石 *Jurusania cylindrica* Krylov, 1963

同义名:

1963 *Jurusania cylindrica* Krylov, Krylov I N, Tr. Geol. Inst. Akad., Nauk SSSR, 69: p. 133.

2008 *Jurusania cylindrica* Krylov, 钱迈平, 汪迎平, 阎永奎, 华北古陆

3.6 *Jurusania cylindrica* Krylov, 1963

Systematics:

Class: Columellati Cao et Yuan, 2006
Order: Ramificolumllati Cao et Yuan, 2006
Family: Kussiellaaceae Raaben, 1969
Genus: *Jurusania* Krylov, 1963
Species: *Jurusania cylindrica* Krylov, 1963

Content:

1963 *Jurusania cylindrica* Krylov, Krylov I N, Tr. Geol. Inst. Akad., Nauk SSSR, 69: p. 133.

2008 *Jurusania cylindrica* Krylov, Qian M P, Wang Y P and Yan Y K, Neoproterozoic Biota in Southeastern Margin of North China, Geological Publishing House, Beijing: p. 91, plate XXVIII: fig. 5.

Description: Parallel cylindrical or subcylindrical columns, rounded to oval or rounded polygonal transverse sections, with partly walled, partly

东南缘新元古代生物群,地质出版社,91 页,图版 XXVIII:图 5。

描述:圆柱-次圆柱状叠层体,不增宽平行分枝,分枝稀少。柱体直径大多 3—4 cm,高超过 20 cm,平行排列,各柱体间距大多 3—5 cm。柱体表面局部有壁,局部发育清晰的檐,一些檐在柱体边缘下垂可超过 1 cm。柱体包裹一层 4—5 mm 厚的泥晶质外套。基本层半球穹形至陡峭穹形,在柱体中部较厚,最厚处可达 1.0—1.5 mm,向边缘变薄下弯。模糊带状微结构。

分布及层位:神农架林区官门山,中元古代石槽河组;江苏-安徽北部,新元古代九顶山组、四顶山组;俄罗斯南乌拉尔,新元古代卡塔夫组(Katav Formation)等。

bear downward directed peaks and overhanging laminae, sometime more than 1 cm long, covered with a 4—5 mm micritic envelope. The columns are 3—5 cm apart, mostly 3—4 cm in diameter and up to more than 20 cm high, rarely no widening parallel braching. Lamina profile varies from simply arch to steep dome, usually thickened in the central part, up to 1.0—1.5 mm thick, with indistinct banding microstructures.

Locality and horizon: Mesoproterozoic Shicaohe Formation, Guanmenshan, Shennongjia Forestry District, Central China; Neoproterozoic Jiudingshan and Sidingshan formations, northern Jiangsu and Anhui provinces; Neoproterozoic Katav Formation, southern Urals, Russia etc.

图 3.6.1 （摄影：钱迈平）　　　　　　　　Fig. 3.6.1 （Photograph by Qian Maiping）

圆柱朱鲁莎叠层石（*Jurusania cylindrica*）纵断面。
化石层位： 中元古代石槽河组
化石地点： 神农架林区官门山公路边

Jurusania cylindrica, longitudinal section.
Horizon: Mesoproterozoic Shicaohe Formation
Locality: Guanmenshan highway, Shennongjia Forestry District, Central China

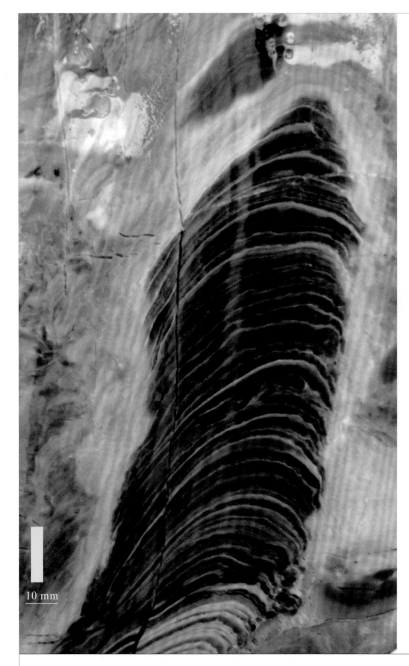

图 3.6.2 （摄影：钱迈平）

Fig. 3.6.2 （Photograph by Qian Maiping）

圆柱朱鲁莎叠层石（*Jurusania cylindrica*）纵切抛光面，显示其柱体在水流中生长的特征，即基本层拱度在迎水流方向陡，伸出的檐因迎着水流而发生卷曲；在顺水流方向缓，伸出的檐因顺着水流而拉长。

A polished longitudinal section of *Jurusania cylindrica*, showing the arched laminae of a stromatolite are steep slope toward and gentle slope parallel to the palaeocurrent direction.

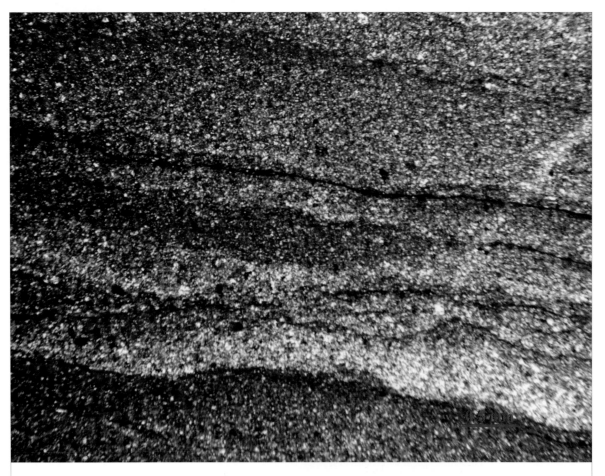

图3.6.3 （摄影：马雪）　　　　　　　　　Fig. 3.6.3 （Photograph by Ma Xue）

圆柱朱鲁莎叠层石（*Jurusania cylindrica*）的基本层纵切面微结构，显示弥散带状的微生物席化石结构。

The microstructure of *Jurusania cylindrical* appears indistinct bands of fossilized microbial mat.

3.7 地窖印卓尔叠层石
(*Inzeria intia* Walter, 1972)

分类:
柱叠层石纲 Columellati Cao et Yuan, 2006
  分枝柱叠层石目 Ramificolumllati Cao et Yuan, 2006
    印卓尔叠层石科 Inzeriaaceae Cao et Yuan, 2006
      印卓尔叠层石属 *Inzeria* Krylov, 1962
        地窖印卓尔叠层石 *Inzeria intia* Walter, 1972

同义名:

1972 *Inzeria intia*, Walter M R, Special papers in palaeontology, No. 11, published by The Palaeontological Association London, pp. 140—147, plate 3: figs. 1—5; plate 20: figs. 4—5; plates 21—23; text-figs. 7: 40—43.

3.7 *Inzeria intia* Walter, 1972

Systematics:
Class: Columellati Cao et Yuan, 2006
Order: Ramificolumllati Cao et Yuan, 2006
Family: Inzeriaaceae Cao et Yuan, 2006
Genus: *Inzeria* Krylov, 1962
Species: *Inzeria intia* Walter, 1972

Content:

1972 *Inzeria intia*, Walter M R, Special papers in palaeontology, No. 11, published by The Palaeontological Association London, pp. 140—147, plate 3: figs. 1—5; plate 20: figs. 4—5; plates 21—23; text-figs. 7: 40—43.

1985 *Inzeria intia*, Cao R J et al., Memoirs of Nanjing Institute of Geology and Palaeontology, No. 21, p. 24, plate Ⅰ: figs. 2, 3.

Description: No widening parallel or widening parallel branching

1985 *Inzeria intia*,曹瑞骥等,中国科学院南京地质古生物研究所集刊,第21号:24页,图版Ⅰ:图2,3。

描述:次圆柱-块茎状叠层体,不增宽或增宽平行分枝,单个柱体宽度变化大,3—15 cm不等,高大于30 cm。各柱体相互平行排列,间距大多0.4—1 cm不等,侧表面发育檐、横肋及连层,局部有壁,大多具壁龛式芽状分叉,芽宽1—3 cm,高3—4 cm,顶端变尖。基本层在柱体较宽部位呈平缓穹状或箱状,其中一部分延至柱体边缘向下弯,局部遮盖柱体侧表面,而另一部分延至柱体边缘终止;在柱体较窄部位凸度较大,呈半球穹状。不规则条带状微结构。

分布及层位:神农架林区神农顶-野马河,中元古代矿石山组、温水河组;安徽灵璧陇山,新元古代九顶山组;澳大利亚北领地(Northern Territory)爱丽斯泉(Allice Springs),新元古代苦泉组(Bitter Springs Formation)等。

subcylindrical or tuberous columns, frequently with niches containing projections. The columns in vertically parallel arrangement, mostly 0.4—1.0 cm apart, are more than 30 cm high, with numerous projections, overhanging peaks, short cornices and links or bridges. The projections are mostly 1—3 cm in diameter, 3—4 cm high. The diameter of a column varies greatly between 3 and 15 cm, with rounded, rounded polygonal or slightly lobate transverse sections. Laminae are always gently convex, varying in shape from continuously curved domes to very low, obtuse cones, with irregular banding microstructures.

Locality and horizon: Mesoproterozoic Kuangshishan and Wenshuihe formations, Shennongding-Yemahe, Shennongjia Forestry District, Central China; Neoproterozoic Jiudingshan Formation, Longshan, Lingbi County, Anhui Province; Neoproterozoic Bitter Springs Formation, Allice Springs, Northern Territory, Australia etc.

图 3.7.1 （摄影：钱迈平） Fig. 3.7.1 （Photograph by Qian Maiping）

地窖印卓尔叠层石（*Inzeria intia*）纵断面。
化石层位：中元古代矿石山组
化石地点：神农架林区神农顶－凉风垭公路边

Inzeria intia, longitudinal section.
Horizon: Mesoproterozoic Kuangshishan Formation
Locality: Shennongding-Liangfengya highway, Shennongjia Forestry District, Central China

图 3.7.2 （摄影：钱迈平） Fig. 3.7.2. (Photograph by Qian Maiping)

地窖印卓尔叠层石（*Inzeria intia*）纵断面。
化石层位：中元古代温水河组
化石地点：神农架林区野马河－鸭子口公路边

Inzeria intia, longitudinal section.
Horizon: Mesoproterozoic Wenshuihe Formation
Locality: Yemahe-Yazikou highway, Shennongjia Forestry District, Central China

图 3.7.3 （摄影：钱迈平）　　　　　　　　Fig. 3.7.3 （Photograph by Qian Maiping）

地窖印卓尔叠层石（*Inzeria intia*）横断面。
化石层位：中元古代温水河组
化石地点：神农架林区野马河－鸭子口公路边

Inzeria intia, transverse section.
Horizon: Mesoproterozoic Wenshuihe Formation
Locality: Yemahe-Yazikou highway, Shennongjia Forestry District, Central China

3.8 瘤通古斯叠层石
(*Tungussia nodosa* Semikhatov, 1962)

分类:
柱叠层石纲 Columellati Cao et Yuan, 2006
  分枝柱叠层石目 Ramificolumllati Cao et Yuan, 2006
    通古斯叠层石科 Tungussiaaceae Raaben, 1969
      通古斯叠层石属 *Tungussia* Semikhatov, 1962
        瘤通古斯叠层石 *Tungussia nodosa* Semikhatov, 1962

同义名:
1962 *Tungussia nodosa* Semikhatov, Semikhatov M A, Academy of Sciences of the SSSR, Geol. Inst. transaction, pp. 205—207; plates Ⅵ, Ⅶ. figs. 1, 2.

描述: 块茎-次圆柱状叠层体,多种方式分枝,以强烈散开式分枝为

3.8 *Tungussia nodosa* Semikhatov, 1962

Systematics:
Class: Columellati Cao et Yuan, 2006
Order: Ramificolumllati Cao et Yuan, 2006
Family: Tungussiaaceae Raaben, 1969
Genus: *Tungussia* Semikhatov, 1962
Species: *Tungussia nodosa* Semikhatov, 1962

Content:
1962 *Tungussia nodosa* Semikhatov, Semikhatov M A, Academy of Sciences of the SSSR, Geol. Inst. transaction, pp. 205—207; plates Ⅵ, Ⅶ. figs. 1, 2.

Description: Tuberous or subcylindrical columns, especially characteristic horizontal or subhorizontal to vertical columns, with frequent, multiple, markedly divergent branching, vary in diameter from 2—4 cm at the base to 15 cm in the upper part. Lamination is distinct but

主,以水平分枝再转垂直向上生长为特征,有时多个子柱体可从一个母柱体的同一部位分出。柱体直径多变,基部宽2—4 cm,往上急剧变粗可达15 cm。基本层凸度、继承性及对称性多变,常在柱体边缘变薄下弯,遮盖柱体表面,形成明显的多层壁。柱体侧表面光滑或具小檐,偶见连层。基本层由均匀的粗晶碳酸盐矿物构成,大多1—2 mm 厚。

分布及层位:神农架林区神农顶-凉风垭,中元古代矿石山组;俄罗斯西伯利亚东北部土鲁汗河(Turukhan River)流域,中元古代苏霍通古斯组(Sukhotungussia Formation)。

changing in form, often deeply bulging or asymmetrical, usually thinned and curved downward in margin to form multiple wall or overhanging peaks, rarely links or bridges. Laminae are constituted of uniformly granular large crystals of dolomite, mostly 1—2 mm thick.

Locality and horizon: Mesoproterozoic Kuangshishan Formations, Shennongding-Yemahe, Shennongjia Forestry District, Central China; Mesoproterozoic Sukhotungussia Formation, Turukhan River region, northeastern Siberia, Russia etc.

图 3.8.1 （摄影：钱迈平）　　　　Fig. 3.8.1 （Photograph by Qian Maiping）

瘤通古斯叠层石(Tungussia nodosa)横断面。
化石层位：中元古代矿石山组
化石地点：神农架林区神农顶－凉风垭公路边

Tungussia nodosa, transverse section.
Horizon: Mesoproterozoic Kuangshishan Formation
Locality: Shennongding-Liangfengya highway, Shennongjia Forestry District, Central China

3.9 贝加尔贝加尔叠层石
(*Baicalia baicalica* (Maslov) Krylov, 1963)

分类:
柱叠层石纲 Columellati Cao et Yuan, 2006
　分枝柱叠层石目 Ramificolumllati Cao et Yuan, 2006
　　通古斯叠层石科 Tungussiaaceae Raaben, 1969
　　　贝加尔叠层石属 *Baicalia* Krylov, 1962
　　　　贝加尔贝加尔叠层石 *Baicalia baicalica* (Maslov) Krylov, 1963
同义名:
　1937 *Collenia baicalica* Maslov, Maslov V P, Problemy Paleont., 2—3: p. 287; plate 4: figs. 2, 3; text-fig. 8.
　1937 *Collenia baicalica* Maslov, Maslov V P, Problemy Paleont., 2—3: p. 333; plate 2: fig. 1; text-fig.1.

3.9 *Baicalia baicalica* (Maslov) Krylov, 1963

Systematics:
Class: Columellati Cao et Yuan, 2006
Order: Ramificolumllati Cao et Yuan, 2006
Family: Tungussiaaceae Raaben, 1969
Genus: *Baicalia* Krylov, 1962
Species: *Baicalia baicalica* (Maslov) Krylov, 1963
Content:
　1937 *Collenia baicalica* Maslov, Maslov V P, Problemy Paleont., 2—3: p. 287; plate 4: figs. 2, 3; text-fig. 8.
　1937 *Collenia baicalica* Maslov, Maslov V P, Problemy Paleont., 2—3: p. 333; plate 2: fig. 1; text-fig.1.
　1960 *Collenia baicalica* Maslov, Krylov I N, Dokl. Akad. Nauk SSSR, 132 (4): p. 896; text-fig.1.
　1960 *Collenia baicalica* Maslov, Semikhatov M A, Dokl. Akad. Nauk SSSR, 135 (6): pp. 1480—1481; text-fig.1: a, b; text-fig.4: a.

1960 *Collenia baicalica* Maslov, Krylov I N, Dokl. Akad. Nauk SSSR, 132 (4): p. 896; text-fig.1.

1960 *Collenia baicalica* Maslov, Semikhatov M A, Dokl. Akad. Nauk SSSR, 135 (6): pp. 1480－1481; text-fig.1: a, b; text-fig.4: a.

1963 *Baicalia baicalica* (Maslov) Krylov, Krylov I N, Akad. Nauk SSSR, Geol. Inst. Trudy, 69: pp. 64－70; text-figs. 18－20, plates Ⅶ－Ⅺ.

1974 *Baicalia baicalica* (Maslov) Krylov,曹瑞骥,梁玉左,中国科学院南京地质古生物研究所集刊,第5号,图版Ⅷ,图1。

1985 *Baicalia baicalica* (Maslov) Krylov,李钦仲,杨应章,贾金昌,华北地台南缘陕西部分晚前寒武纪地层研究,西安交通大学出版社,112－113页,图版9:图1。

2008 *Baicalia baicalica* (Maslov) Krylov,钱迈平,汪迎平,阎永奎,华北古陆东南缘新元古代生物群,地质出版社,93页,图版ⅩⅩⅨ:图4。

描述:块茎状叠层体,大小不一,基部收缩,向上膨胀,直径1.5－18 cm不等,高7－24 cm不等。散开分枝,分枝处明显收缩,分出子柱体基部窄,向上增宽。柱体侧表面不平整,基本层平缓穹状,常延伸至柱体外形成檐,或在柱体边缘叠覆,无壁。条带状微结构。

分布及层位:神农架林区官门山,中元古代石槽河组;辽宁大连金州大李家屯,新元古代十三里台组;江苏徐州贾汪,新元古代魏集组;俄罗斯西伯利亚贝加尔湖地区,新元古代尤伦图组(Uluntui Formation)等。

1963 *Baicalia baicalica* (Maslov) Krylov, Krylov I N, Akad. Nauk SSSR, Geol. Inst. Trudy, 69: pp. 64－70; text-figs. 18－20, plates Ⅶ－Ⅺ.

1974 *Baicalia baicalica* (Maslov) Krylov, Cao R-J and Liang Y Z, Memoirs of Nanjing Institute of Geology and Palaeontology, No. 5, plate Ⅶ: fig. 1.

1985 *Baicalia baicalica* (Maslov) Krylov, Li Q Z, Yang Y Z and Jia J C, The Study of Late Precambrian Strata in the Southern Margin of the North China Platform (Part of Shaanxi Province), Xi'an Jiaotong University Press, Xi'an: pp. 112－113, plate 9: fig. 1.

2008 *Baicalia baicalica* (Maslov) Krylov, Qian M P, Wang Y P and Yan Y K, Neoproterozoic Biota in the Southeastern Margin of the North China Paleocontinent, Geological Publishing House, Beijing: p. 93, plate ⅩⅩⅨ: fig. 4.

Description: Divergent branching tuberous columns vary sharp in diameter from about 1.5 cm at the base to about 18 cm in the upper part, 7－24 cm high. The lateral surface of the column is uneven, non-wall, with discrete cornices or overlapping margins formed by laminae. The laminae appear mostly gently convex, with banding microstructures.

Locality and horizon: Mesoproterozoic Shicaohe Formation, Guanmenshan, Shennongjia Forestry District, Central China; Neoproterozoic Shisanlitai Formation, Dalijiatun, Jinzhou District, Dalian City, Liaoning Province; Neoproterozoic Weiji Formation, Jiawang District, Xuzhou City, Jiangsu Province; Neoproterozoic Uluntui Formation, Baikal region, Siberia, Russia etc.

图 3.9.1 （摄影：钱迈平） Fig. 3.9.1 （Photograph by Qian Maiping）

这是贝加尔贝加尔叠层石（*Baicalia baicalica*）纵断面。
化石层位：中元古代石槽河组
化石地点：神农架林区官门山攀岩基地

Baicalia baicalica, longitudinal section.
Horizon: Mesoproterozoic Shicaohe Formation
Locality: Guanmenshan, Shennongjia Forestry District, Central China

3.10 育卡贝加尔叠层石
(*Baicalia unca* Semikhatov, 1962)

分类:

柱叠层石纲 Columellati Cao et Yuan, 2006
 分枝柱叠层石目 Ramificolumllati Cao et Yuan, 2006
  通古斯叠层石科 Tungussiaaceae Raaben, 1969
   贝加尔叠层石属 *Baicalia* Krylov, 1962
    育卡贝加尔叠层石 *Baicalia unca* Semikhatov, 1962

同义名:

1962 *Baicalia unca* Semikhatov, Semikhatov M A, Academy of Sciences of the SSSR, Geol. Inst. Transaction, pp. 202—203; plate Ⅳ: figs. 1—3.

3.10 *Baicalia unca* Semikhatov, 1962

Systematics:

Class: Columellati Cao et Yuan, 2006
Order: Ramificolumllati Cao et Yuan, 2006
Family: Tungussiaaceae Raaben, 1969
Genus: *Baicalia* Krylov, 1962
Species: *Baicalia unca* Semikhatov, 1962

Content:

1962 *Baicalia unca* Semikhatov, Semikhatov M A, Academy of Sciences of the SSSR, Geol. Inst. transaction, pp. 202—203; plate Ⅳ: figs. 1—3.

1986 *Baicalia aimica* Nuzhnov, Cao R J, The Bureau of Geology and Mineral Resources of Jilin Province, Paleontological Atlas of Jilin China, Jilin Science and Technology Publishing House, Changchun: p. 581, plate 270: fig. 1.

1986 *Baicalia aimica* Nuzhnov, 曹瑞骥, 吉林省地质矿产局主编, 吉林省古生物图册, 吉林科学技术出版社, 581页: 图版270: 图1。

描述: 块茎－次圆柱状叠层体, 直径1—5 cm不等, 大多高5 cm。强烈散开式二分枝, 分枝处明显或不明显收缩, 分出子柱体基部窄, 向上增宽或强烈增宽。柱体侧表面不平整, 基本层大多半球穹状, 常延伸至柱体外较长, 形成明显的檐, 偶有连层, 无壁。断续线状微结构。

分布及层位: 神农架林区神农顶和野马河, 中元古代乱石沟组和野马河组; 吉林白山浑江清沟子, 新元古代八道江组;

Description: Tuberous or subcylindrical columns, with markedly Y-shaped diverging branches, vary in diameter from about 1 cm at the base to about 5 cm in the upper part, mostly 5 cm high. The lateral surface of the column is uneven, non-wall, with discrete cornices, rarely links. The laminae are generally moderately to steeply convex, with discontinuous banding microstructures.

Locality and horizon: Mesoproterozoic Luanshigou and Yemahe formations, Shennongding and Yemahe; Shennongjia Forestry District, Central China; Neoproterozoic Badaojiang Formation, Qinggouzi, Hunjiang District, Baishan City, Jilin Province etc.

图 3.10.1 （摄影：钱迈平）　　　　　　　　Fig. 3.10.1　Photograph by Qian Maiping

育卡贝加尔叠层石（*Baicalia unca*）横断面。
化石层位：中元古代野马河组
化石地点：神农架林区野马河－鸭子口公路边

Baicalia unca, transverse section.
Horizon: Mesoproterozoic Yemahe Formation
Locality: Yemahe-Yazikou highway, Shennongjia Forestry District, Central China

3.11 奥姆泰尼奥姆泰尼叠层石
(*Omachtenia omachtensis* Nuzhnov, 1967)

分类:

层柱叠层石纲 Stratcolumellati Cao et Yuan, 2006
　奥姆泰尼叠层石科 Omachteniaaceae Koniushkov, 1978
　　奥姆泰尼叠层石属 *Omachtenia* Nuzhnov, 1967
　　　奥姆泰尼奥姆泰尼叠层石 *Omachtenia omachtensis* Nuzhnov, 1967

同义名:

1967 *Omachtenia omachtensis*, Nuzhnov S V, Geol. Inst. Yahutsk, Filial Sibirsk. Otdel. Akad. Nauk. SSSR, Moscow, 160pp.

1988 *Omachtenia omachtensis*, Walter M R, et al., Alcheringa, 12: pp. 79—106.

3.11 *Omachtenia omachtensis* Nuzhnov, 1967

Systematics:

Class: Stratcolumellati Cao et Yuan, 2006
Family: Omachteniaaceae Koniushkov, 1978
Genus: *Omachtenia* Nuzhnov, 1967
Species: *Omachtenia omachtensis* Nuzhnov, 1967

Content:

1967 *Omachtenia omachtensis*, Nuzhnov S V, Geol. Inst. Yahutsk, Filial Sibirsk. Otdel. Akad. Nauk. SSSR, Moscow, 160pp.

1988 *Omachtenia omachtensis*, Walter M R, et al., Alcheringa, 12: pp. 79—106.

Description: Stratiform-columnar structures, with no widening parallel branching subcylindrical columns in vertically tightly parallel arrangement, almost without interspace, mostly 4 cm in diameter, non-wall, with discrete cornices and links. There are 2 or 3 upward branches created from

描述:次圆柱－层状叠层体,不增宽平行分枝,紧密排列。母柱体大多宽4 cm,向上分出2—3个子柱体,子柱体宽1—2 cm,无壁,发育大量檐和连层。基本层大多平缓穹状,局部半球至强烈凸起穹状。条带状微结构。

分布及层位:神农架林区官门山,中元古代石槽河组;天津蓟县下营大红峪沟口,古元古代大红峪组上部;俄罗斯东西伯利亚乌丘尔河(Uchur River)流域,中元古代奥姆泰尼组(Omachta Formation)等。

the top of a column, and the branches are usually 1—2 cm in diameter. The laminae are generally gently convex and partially hemispherical to steeply convex, with banding microstructures.

Locality and horizon: Mesoproterozoic Shicaohe Formation, Guanmenshan, Shennongjia Forestry District, Central China; Paleoproterozoic Dahongyu Formation, Dahongyu, Jixian, Tianjin City; Mesoproterozoic Omachta Formation, Uchur River region, eastern Siberia, Russia etc.

图 3.11.1 （摄影：钱迈平）　　　　　　　　　　Fig. 3.11.1 （Photograph by Qian Maiping）

奥姆泰尼奥姆泰尼叠层石（*Omachtenia omachtensis*）纵断面。
化石层位： 中元古代石槽河组
化石地点： 神农架林区官门山攀岩基地

Omachtenia omachtensis, longitudinal section.
Horizon: Mesoproterozoic Shicaohe Formation
Locality: Guanmenshan, Shennongjia Forestry District, Central China

3.12 波层叠层石
(*Stratifera undata* Komar, 1966)

分类:

层叠层石纲 Stratiformati
　层叠层石科 Stratiferaacea Raaben and Sinha, 1989
　　层叠层石属 *Stratifera* Korolyuk, 1956
　　　波层叠层石 *Stratifera undata* Komar, 1966

同义名:

1966 *Stratifera undata* Komar, Komar V A, Tr. Geol. Inst. Akad., Nauk SSSR, Akad. Nauk SSSR, 154: pp. 69—70.

1986 *Stratifera undata* Komar, Fairchild T R and Subacius S M R, Precambrian Research, 33: p. 327, fig. 3.

2015 *Stratifera undata* Komar, 钱迈平等, 浙东石浦组叠层石-虫管生物礁灰岩地层的年龄与沉积相. 地层学杂志, 第39卷, 第2期: 176页, 图5.

描述: 层状叠层体, 基本层波状起伏, 起伏幅度和频率多变。团块及条带状微结构。

分布及层位: 神农架林区, 中元古代神农架群碳酸盐岩地层广泛分布。

3.12 *Stratifera undata* Komar, 1966

Systematics:

Class: Stratiformati
Family: Stratiferaacea Raaben and Sinha, 1989
Genus: *Stratifera* Korolyuk, 1956
Species: *Stratifera undata* Komar, 1966

Content:

1966 *Stratifera undata* Komar, Komar V A, Tr. Geol. Inst. Akad., Nauk SSSR, Akad. Nauk SSSR, 154: pp. 69—70.

1986 *Stratifera undata* Komar, Fairchild T R and Subacius S M R, Precambrian Research, 33: p. 327, fig. 3.

2015 *Stratifera undata* Komar, Qian M P, et al., Journal of Stratigraphy, 39(2): pp. 169—187, fig. 5.

Description: Stratiform structures, with laterally linked flattened, moderated to disordered undatary laminae, lumpy and banding microstructures.

Locality and horizon: Mesoproterozoic carbonate rock formations in Shennongjia Forestry District, Central China.

图3.12.1 （摄影：钱迈平）　　　　　　　　　Fig. 3.12.1 （Photograph by Qian Maiping）

波层叠层石(*Stratifera undata*)。
化石层位：中元古代乱石沟组
化石地点：神农架林区神农谷－神农顶公路边

Stratifera undata.
Horizon: Mesoproterozoic Luanshigou Formation
Locality: Shennonggu-Shennongding highway, Shennongjia Forestry District, Central China

图3.12.2 （摄影：钱迈平）　　　　　　　　Fig. 3.12.2 （Photograph by Qian Maiping）

波层叠层石（*Stratifera undata*）。
化石层位：中元古代大窝坑组
化石地点：神农架林区神农顶－凉风垭公路边

Stratifera undata.
Horizon: Mesoproterozoic Dawokeng Formation
Locality: Shennongding-Liangfengya highway, Shennongjia Forestry District, Central China

图 3.12.3
(摄影:钱迈平)

Fig. 3.12.3 (Photograph by Qian Maiping)

中元古代大窝坑组的波层叠层石(*Stratifera undata*)基本层纵切面。

A polished longitudinal section of *Stratifera undata*. from Mesoproterozoic Dawokeng Formation.

图 3.12.4
(摄影:马雪)

Fig. 3.12.4 (Photograph by Ma Xue)

波层叠层石(*Stratifera undata*)基本层纵切面微结构,显示团块及丝状微生物席化石结构。

The microstructure of *Stratifera undata* shows lumpiness and wire-like fossilized microbial mat laminae.

结 语

神农架地质公园主要景区沿公路出露的中元古代地层,自下而上包括:乱石沟组、大窝坑组、矿石山组、台子组、野马河组、温水河组和石槽河组等。叠层石生物礁白云岩在碳酸盐台地相沉积地层十分发育,甚至在一些地方保存了10多亿年前形成的绵延数十至上百千米的叠层石大堡礁遗迹。其中以大窝坑组、矿石山组及台子组下部地层的叠层石生物礁白云岩最发育。

叠层石类型主要有:神农架大圆顶叠层石(*Megadomia shennongjiaensis*)、加尔加诺锥叠层石(*Conophyton garganicum*)、树桩圆柱叠层石(*Colonnella cormosa*)、简单包心菜叠层石(*Cryptozoon haplum*)、喀什喀什叠层石(*Kussiella kussiensis*)、圆柱朱鲁莎叠层石(*Jurusania cylindrica*)、地窖印卓尔叠层石(*Inzeria intia*)、瘤通古斯叠层石(*Tungussia nodosa*)、贝加尔贝加尔叠层石(*Baicalia baicalica*)、育卡贝

CONCLUSIONS

The Mesoproterozoic sequences exposed along highways in main tourist spots of Shennongjia UNESCO Global Geopark in an ascending order: Luanshigou, Dawokeng, Kuangshishan, Taizi, Yemahe, Wenshuihe and Shicaohe formations etc. The stromatolite dolostones were well developed in carbonate plateform sedimentary sequences, and even preserved remains of more than one billion years old great barrier reefs made up of stromatolite bioherms stretched several dozens or hundreds kilometers in this district. Most of stromatolite dolostones occurred in Dawokeng Formation, Kuangshishan Formation and lower part of Taizi Formation.

Various stromatolites in Shennongjia UNESCO Global Geopark include *Megadomia shennongjiaensis*, *Conophyton garganicum*, *Colonnella cormosa*, *Cryptozoon haplum*, *Kussiella kussiensis*, *Jurusania cylindrical*, *Inzeria intia*, *Tungussia nodosa*, *Baicalia baicalica*, *Baicalia unca*, *Omachtenia omachtensis* and *Stratifera undata* etc. They were booming in a carbonate plateform off the coast of South China Craton located northern

加尔叠层石(*Baicalia unca*)、奥姆泰尼奥姆泰尼叠层石(*Omachtenia omachtensis*)和波层叠层石(*Stratifera undata*)等。由于当时神农架位于华南古陆,在罗迪尼亚超级大陆(Rodinia Supercontinent)北部,其叠层石主要生长在沿海的碳酸盐台地环境,其组合面貌与相邻的澳大利亚古陆和西伯利亚古陆沿海的叠层石组合面貌相似,其中多数类型是相同的。

  随着海水变深,光照减弱,微生物席光合作用难以维系。同时,适合叠层石生物礁发育的碳酸盐台地也变成了不适合叠层石形成的泥砂质海底。最终在约10亿年前的新元古代初期,神农架的叠层石生物礁消失。

Rodinia Supercontinent, and similar to adjoining Australia Craton and Siberia Craton in stromatolite assemblages during the Mesoproterozoic Era.

With the water getting deeper and deeper, the illumination was weaker and weaker at the bottom, the photosynthesis of microbial mats was more and more difficult to keep on, and meanwhile the carbonate plateform fit for stromatolite forming transformed into mud or sandy seabed unsuited to them. The stromatolites eventually disappeared in Shenongjia region approximately 1,000 billion years ago.

致　谢

本项目野外考察和实验测试期间,得到神农架世界地质公园管理局办公室主任王志先、干部钟权,以及中国地质调查局南京地质调查中心研究员杨祝良、高天山,副研究员陈荣,助理研究员张炜、靳国栋、彭博、郑剑威、段政、张翔、洪文涛,高级工程师周效华、余明刚、朱清波、李春海、姜杨、蒋仁、于俊杰,湖北省地质调查研究院教授级高级工程师邱艳生的大力协助,在此谨表诚挚的谢意。

ACKNOWLEDGEMENTS

We would like to give our special thank to the Office Director Wang Zhixian, Official Zhong Quan of the Administration Bureau of Shennongjia UNESCO Global Geopark; Professor Yang Zhuliang, Gao Tianshan, Associate Professor Chen Rong, Research Assistant Zhang Wei, Jin Guodong, Peng Bo, Zheng Jianwei, Duan Zheng, Zhang Xiang, Hong Wentao, Senior Engineer Zhou Xiaohua, Yu Minggang, Zhu Qingbo, Li Chunhai, Jiang Yang, Jiang Ren, Yu Junjie from Nanjing Center of China Geological Survey and Professorate Senior Engineer Qiu Yansheng from Geological Survey Institute of Hubei Province for giving us their full cooperation during the field survey and experimental measurements.

参 考 文 献
REFERENCES

曹瑞骥,1986.吉林地区叠层石类 [M] //吉林省地质矿产局.吉林省古生物图册.长春:吉林科学技术出版社:581-585.

曹瑞骥,梁玉左,1974.从藻化石和叠层石论中国震旦系划分和对比 [J].中国科学院南京地质古生物研究所集刊(5):1-16.

曹瑞骥,袁训来,2006.叠层石 [M].合肥:中国科学技术大学出版社:1-383.

曹瑞骥,赵文杰,1977.辽东震旦亚界的叠层石组合及其地层意义 [M] // 中国科学院铁矿地质学术会议论文选集:地层古生物.北京:科学出版社:61-86.

曹瑞骥,赵文杰,夏广胜,1985.安徽北部晚前寒武纪叠层石 [J].中国科学院南京地质古生物研究所集刊,21:1-84.

国家地质总局天津地质矿产研究所,中国科学院南京地质古生物研究所,内蒙古自治区地质矿产局,1979.蓟县震旦亚界叠层石研究 [M].北京:地质出版社:1-94.

李怀坤,张传林,相振群,等,2013.扬子克拉通神农架群锆石和斜锆石U-Pb年代学及其构造意义 [J].岩石学报,29(2):673-697.

李铨,冷坚,1991.前寒武地质研究:神农架上前寒武系 [M].天津:天津科学技术出版社:1-503.

李钦仲,杨应章,贾金昌,1985.华北地台南缘(陕西部分)晚前寒武纪地层研究 [M].西安:西安交通大学出版社:104-158.

梁玉左,1980.我国北方燕辽地区震旦亚界的叠层石和核形石及其地层意义.地层古生物论文集,8(1):28-29.

缪长泉,1993.新疆昆仑山和阿尔金山前寒武系及叠层石 [M].乌鲁木齐:新疆科技卫生出版社:86-87.

钱迈平,姜杨,余明刚,2009.苏皖北部新元古代宏体碳质化石 [J].古生物学报,48(1):73-88.

钱迈平,汪迎平,阎永奎,2008.华北古陆东南缘新元古代生物群 [M].北京:地质出版社:75-94.

钱迈平,张宗言,姜杨,等,2012.中国东南部新元古代冰碛岩地层 [J].地层学杂志,36(3):587-589.

钱迈平,张宗言,余明刚,等,2015.浙东石浦组叠层石-虫管生物礁灰岩地层的年龄与沉积相 [J].地层学杂志,39(2):169-187.

张录易,邱树玉,曹瑞骥,1982.叠层石 [M]// 西安地质矿产研究所.西北地区古生物:陕甘宁分册(一).北京:地质出版社:347-372.

Bell E A, Boehnike P, Harrison T M, et al., 2015. Potentially biogenic carbon preserved in a 4.1 billion-year-old zircon[J]. Proc. Natl. Acad. Sci. U.S.A., 112(47): 14518-14521.

Black M, 1933. The algal sedimentation of Andros Island Bahamas[J]. Philosophical

Transactions of the Royal Society (London) Series B: Biological Science, 222: 165-192.

Cloud P E, Semikhatov M A, 1969. Proterozoic stromatolite zonation[J]. American Journal of Science, 267: 1017-1061.

Djokic T, Van Kranendonk M J, Campbell K A, et al., 2017. Earliest signs of life on land preserved in ca. 3.5 Ga hot spring deposits[J]. Nature communications, 8: Article number 15263.

El Albani A, Bengtson S, Canfield D E, et al., 2010. Large colonial organisms with coordinated growth in oxygenated environments 2.1 Gyr ago[J]. Nature, 466 (7302): 100-104.

Farías M E, Rascovan N, Toneatti D M, et al., 2013. The discovery of stromatolites developing at 3,570 m above sea level in a high-altitude volcanic lake Socompa, Argentinean Andes[J]. PLoS One, 8 (1): e53497.

Glaessner M F, Preiss W V, Walter M R, 1969. Precambrian columnar stromatolites in Australia: morphological and stratigraphic analysis[J]. Science, 164 (3883): 1056-1058.

Goodge J W, Vervoort J D, Fanning C M, et al., 2008. A positive test of East Antarctica-Laurentia juxtaposition within the Rodinia Supercontinent[J]. Science, 321 (5886): 235-240.

Hall J, 1883. *Cryptozoön*, n.g.; *Cryptozoön proliferum*, nsp. New York State Museum of Natural History, 36th Annual Report of the Trustees, plate 6.

Hoffman P F, Kaufman A J, Halverson G P, et al.,1998. A Neoproterozoic snowball earth[J]. Science, 281 (5381): 1342-1346.

Hoffman P F, Schrag D P, 2000. Snowball earth[J]. Scientific American, 282 (1): 68-75.

Hoffman P F, Schrag D P. 2002. The snowball Earth hypothesis: Testing the limits of global change[J]. Terra Nova, 14 (3): 129-155.

Hofmann H J, 1969. Attributes of stromatolites[J]. Geological Survey of Canada Paper: 68-69.

Kalkowsky E, 1908. Oölith und Stromatolith im norddeutschen Buntsandstein[J]. Zeitschrift der Deutschen geologischen Gesellschaft, 60: 68-125.

Keller S C, Bessell M S, Frebel A, et al., 2014. A single low-energy, iron-poor supernova as the source of metals in the star SMSS J031300.36 − 670839.3[J]. Nature. 506(7489): 463-466.

Komar V A, 1964. Riphean Columnar stromatolites from the North Siberia platform [J]. Academy of Sciences of the USSR Reports, 154 (6): 84-105 [in Russian]; English translation by American Geological Institute, 1965, Academy of Science, USSR Doklady,154: 69-70.

Komar V A, 1966. Upper Precambrian stromatolites in the North Siberian platform and their stratigraphic significance[J]. Akad. Nauk SSSR, 154: 69-70 [in Russian].

Komar V A, Raaben M E, Semikhatov M A, 1965. The Riphean *Conophyton* forms in the USSR and their stratigraphic significance[J]. Academy of Sciences of the USSR Reports, 131: 1-72 [in Russian]; English translation by American Geological Institute, 1965, Academy of Science, USSR Doklady, 131: 42-46.

Korolyuk I K, 1959. Undulatory-bedded stromatolites (*Stratifera*) in the Cambrian rocks of southeastern Siberia.

Korolyuk I K, 1960. Stromatolites from the Lower Cambrian and Proterozoic of the Irkutsk Amphitheater[J]. Trudy. Inst. Geol. Razrab Goryuch Iskop, Akad. Nauk SSSR, 1: 35 [in Russian].

Korolyuk I K, 1963. Stromatolites of the Late Precambrian in the Upper Precambrian, Stratigrafiya USSR[M]. Moscow: 479-498 [in Russian].

Krylov I N, 1960. Concerning the development of branching columnar stromatolites in the Riphean of the southern Urals[J]. Doklady Akademii Nauk SSSR, 132 (4): 895-896 [in Russian]; English translation by American Geological Institute, 1960, Academy of Science, USSR Doklady, 132: 515-517.

Krylov I N, 1963. Columnar branching stromatolites of the Riphean deposits of the southern Ural and their significance of the Upper Precambrian[J]. Akad. Nauk SSSR, Geological Institute, Trudy, 69: 60-63, 133 [in Russian].

Li Z X, Bogdanova S V, Collins A S, et al., 2008. Assembly, configuration, and break-up history of Rodinia: A synthesis[J]. Precambrian Research, 160 (1-2): 179-210.

Maslov V P, 1939. An attempt to determine the age of the Ural's barren strata with the aid of stromatolites[J]. Probl Paleontol 5: 297-310.

Long D G F. 1993. The Burgsvik Beds, an Upper Silurian storm generated sand ridge complex in southern Gotland, Sweden[J]. GFF. 115 (4): 299-309.

McMenamin M A S, 1996. Ediacaran biota from Sonora, Mexico[J]. Proceedings of the National Academy of Sciences (USA), 93: 4990-4993.

Nutman A P, Bennett V C, Friend C R L, et al., 2016. Rapid emergence of life shown by discovery of 3,700-million-year-old microbial structures[J]. Nature, 537: 535-538.

Nuzhnov S V, 1967. Rifeyskie otlozheniya yugo-vostoka Sibirskoy platformy (Riphean deposits of the southeast Siberian Platform). Inst. Geol. Yakutsk. Filial Sibirsk. Otdel. Akad. Nauk. SSSR. Moscow, 1-160.

Qian M P, Xing G F, Ma X, et al., 2017. The Mesoproterozoic giant stromatolites from Shennongjia UNESO Global Geopark, Central China [J]. Journal of Geology, 41 (4): 523-528.

Ricardi-Branco F, Caires E, Silva A M, 2006. Gigant Stromatolites field of Santa Rosa de Viterbo, State of São Paulo. Excellent record of the Irati Permian sea coastal environment, Paraná Basin[M]// Winge M, Schobbenhaus C, Berbert-Born M, et al. Sítios geológicos e paleontológicos do Brasil, Vol. II: 1-9.

Semikhatov M A, 1962. Academy of sciences of the SSSR, Geol. Inst. transaction: 202-203.

Som S M, Buick R, Hagadorn J W, et al., 2016. Earth's air pressure 2.7 billion years ago constrained to less than half of modern levels[J]. Nature Geoscience. 9 (6): 448-451.

Shen B, Dong L, Xiao S H, Kowalewski M, 2008. The Avalon Explosion: Evolution of Ediacara Morphospace[J]. Science, 319 (5859): 81-84.

Steele J H, 1825. A description of the Oolitic Formation lately discovered in the county of Saratoga, and state of New York[J]. American Journal of Science, 9: 16-19.

Sumner D Y, 2004. Implications for Neoarchaean ocean chemistry from primary carbonate mineralogy of the Campbellrand-Malmani Platform, South Africa[J]. Sedimentology, 51 (6): 1273-1299.

Walcott C D, 1914. Cambrian geology and paleontology Ⅲ: Precambrian Algonkian algal flora[J]. Smithsonian Miscellaneous Collection, 64: 77-156.

Walter M R, 1972. Stromatolites and the biostratigraphy of the Australian Precambrian and Cambrian[M]//Special Papers in Palaeontology, 11. The Palaeontological Association London: 110-147.

Vally J W, Cavosie A J, Ushikubo T, et al., 2014. Hadean age for a post-magma-ocean zircon confirmed by atom-probe tomography[J]. Nature Geoscience, 7: 219-223.

Walter M R, Krylov I N, Muir M D, 1988. Stromatolites from Middle and Late Proterozoic sequences in McArthur and Georgina Basins and the Mount Isa Province, Australia [J]. Alcheringa, 12: 79-106.

Yuan X L, Chen Z, Xiao S H, et al., 2011. An early Ediacaran assemblage of macroscopic and morphologically differentiated eukaryotes[J]. Nature, 470 (7334): 390-393.

Zhu S X, Zhu M Y, Knoll A H, et al., 2016. Decimetre-scale multicellular eukaryotes from the 1.56-billion-year-old Gaoyuzhuang Formation in North China[J]. Nature Communications 7: 11500. Published online 2016 May 17. doi: 10.1038/ncomms 11500.